Android产品实战从零开始

黄宇健　编著

清华大学出版社
北　京

内容简介

本书包含9个章节，第一章介绍环境搭建以及 Android 基本开发框架；第二章介绍 Android 基本控件；第三到八章分别讲解了六个不同的应用实例；第九章介绍了 Android4. x 的新特性。各个章节的重点如下：第1章 Android 环境搭建、Android 开发框架、ADT 的使用；第2章 四大组件、五大布局、基本控件的使用；第3章 ListView、数据存储、Notification、AppWidget，讲解应用 Timetable；第4章 ExpandableListView、Animation，讲解应用 to-do；第5章 SurfaceView、浮窗，讲解应用 Clock；第6章 调用系统服务、获取系统信息，讲解应用 Easearch；第7章 地图开发、传感器开发、相机开发、Canvas 绘图，讲解应用 MyWhere；第8章 fragment、ViewFlipper、MediaPlayer，讲解应用 YiRstr；第9章 ViewPager、PagerTitleStrip、GridLayout、增强 Notification。

本书讲解详细，并借用案例讲解，有一定的参照性，适合 Android 初学者学习。

图书在版编目(CIP)数据

Android 产品实战从零开始/黄宇健编著. --北京：清华大学出版社，2014
ISBN 978-7-302-37404-6

Ⅰ. ①A… Ⅱ. ①黄… Ⅲ. ①移动终端－应用程序－程序设计 Ⅳ. ①TN929.53

中国版本图书馆 CIP 数据核字(2014)第162967号

责任编辑：魏江江 薛 阳
封面设计：杨 兮
责任校对：焦丽丽
责任印制：刘海龙

出版发行：清华大学出版社
网 址：http://www.tup.com.cn，http://www.wqbook.com
地 址：北京清华大学学研大厦A座 邮 编：100084
社 总 机：010-62770175 邮 购：010-62786544
投稿与读者服务：010-62776969，c-service@tup.tsinghua.edu.cn
质 量 反 馈：010-62772015，zhiliang@tup.tsinghua.edu.cn
印 装 者：北京嘉实印刷有限公司
经 销：全国新华书店
开 本：185mm×260mm **印 张**：19 **字 数**：463千字
版 次：2014年9月第1版 **印 次**：2014年9月第1次印刷
印 数：1～2000
定 价：49.00元

产品编号：060772-01

我是在2010年的时候开始接触Android的。那时候Android吸引我的原因很简单：苹果的体验，诺基亚的价格。后来又在程序员杂志中看到一篇介绍Android开发的文章，非常惊讶地发现Android的开发语言是我最熟悉的Java，于是就开始了Android的开发之路。

Android给我带来的是完全不同于桌面应用、网站的体验，开发Android应用对我来说就是一种乐趣。把这种乐趣和读者们一起分享也是我写这本书的初衷之一。

2012年的夏天我出版了我的第一本Android入门书籍。现在来看，当时写的那本书不够完善，因此才会有写这第二本书的想法。人总是在学习进步的过程中，就算是现在来看刚开始写这本书的半年前，也会发现有些纰漏及不妥的地方，虽然在写后面的章节会慢慢地回过头去修改前面的内容，但总还是觉得不够好，也只能说，如果要磨出一本好书的话，这点时间确实不够。而IT行业却是日新月异的行业，这本书截稿的时候刚好推出了Android 4.3，但是等到出版的时候可能已经进化到Android 5.0了。如果要等万事都完美了再来出版这本书，那恐怕还赶不上更新速度。

站在一个普通读者的角度上来说，对于工具类的开发书籍，我的使用方法是：在电脑上读代码而不是在书里。在电脑上读代码好处很多，在读代码时，Ctrl一下就可以跳到那个方法，跳来跳去非常方便，而书就做不到这一点。因此我尽量少在书中贴不必要的代码，本书配套资料中会有附带本书讲到的所有代码，读者可以放到电脑上读。但是作为技术书籍，代码肯定是会有的，而且还占比较大的篇幅。在书中，我尽量用图来说话，将每个模块都分步讲解，希望广大读者可以接受这种方式。

由于时间匆促，学识有限，书中不足和疏漏之处在所难免，恳请广大读者将意见与建议反馈给我们，以便在后续版本中不断改进和完善。

1. 本书的章节安排

本书一共9个章节，各章的重点分别为：

第1章　Android环境搭建、Android开发框架、ADT的使用。

第2章　四大组件、五大布局、基本控件的使用。

第3章　ListView、数据存储、Notification、AppWidget，讲解应用Timetable。

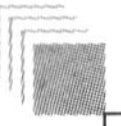

第 4 章　ExpandableListView、Animation，讲解应用 to-do。

第 5 章　SurfaceView、浮窗，讲解应用 Clock。

第 6 章　调用系统服务、获取系统信息，讲解应用 Easearch。

第 7 章　地图开发、传感器开发、相机开发、Canvas 绘图，讲解应用 MyWhere。

第 8 章　fragment、ViewFlipper、MediaPlayer，讲解应用 YiRstr。

第 9 章　ViewPager、PagerTitleStrip、GridLayout、增强 Notification。

本书以“总-分”的形式安排各个章节，先讲解章节中使用的各个知识点，再讲解这些知识点如何运用到本书当中。

2. 本书适合的读者

◇ 略懂 Java 的程序员；

◇ Android 开发的初学者；

◇ 完全没有尝试过 Android 开发的开发人员。

3. 鸣谢

感谢总是不遗余力帮助我的郭老师，总是能给我带来正能量的 Maya，只有数面之缘的 Stan 夫妇，AIESEC 的 Connie 以及 Alex，当然还有老爸老妈，一起玩乐的兄弟们。

黄宇健

2014 年 4 月

目录

CONTENTS

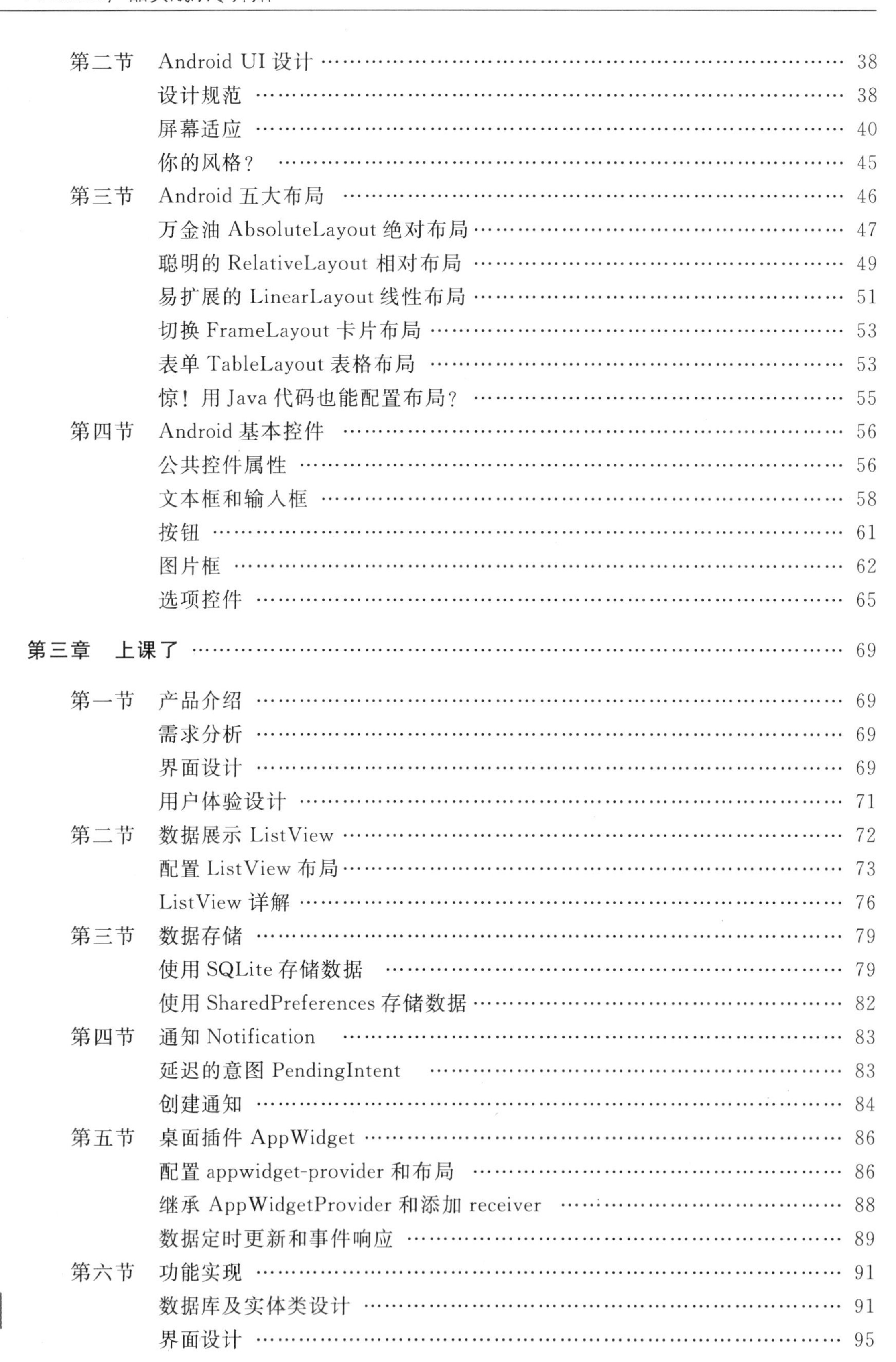

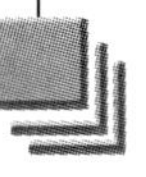

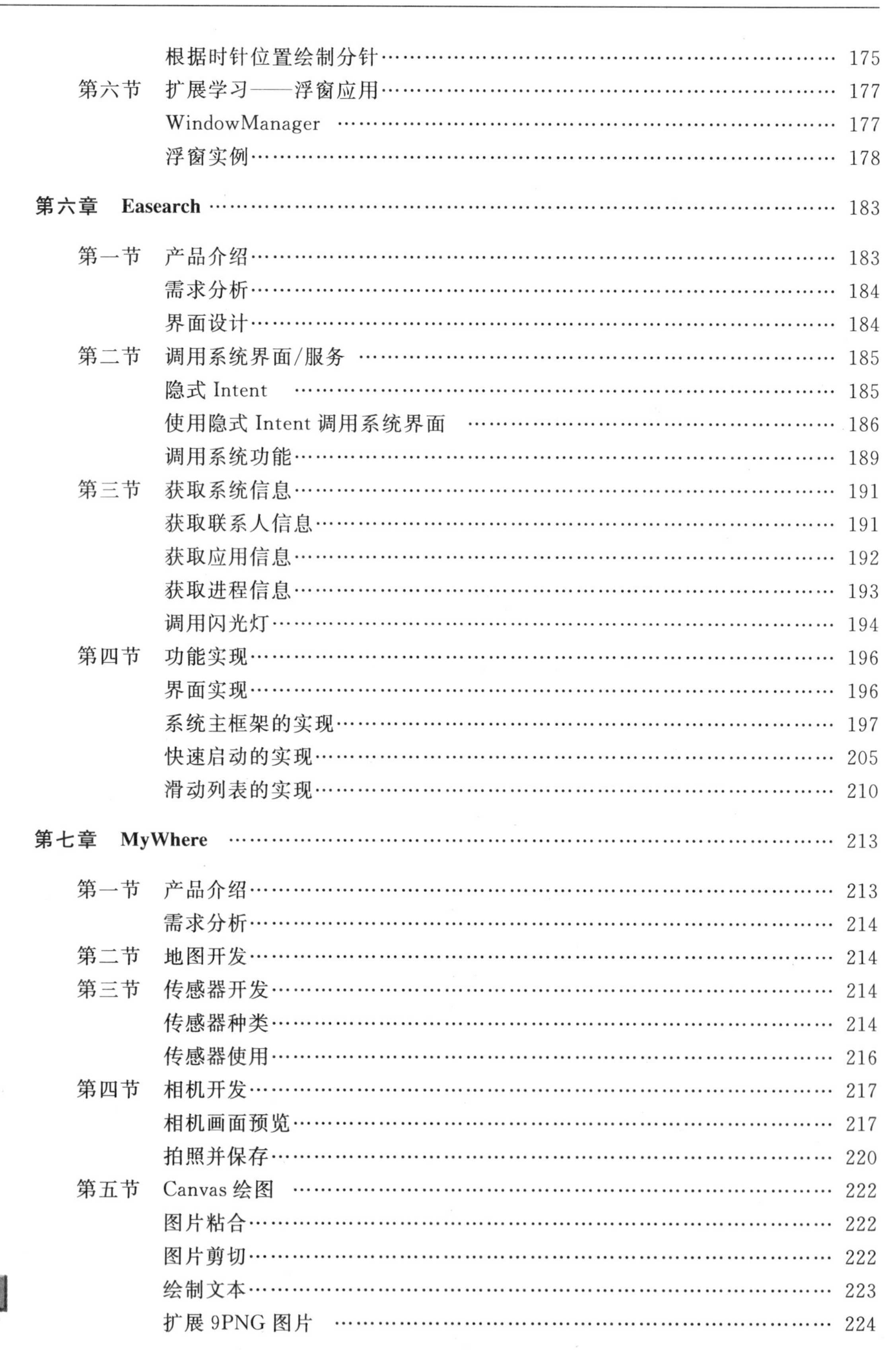

第一章 Android入门

第一节 Android 系统概述

Android 世界百态

Android 是一种基于 Linux 的自由及开放源代码的操作系统，主要适用于便携设备，如智能手机和平板电脑……我知道读者懂得怎么上网百度，所以我不打算把一整段百度百科的文字放到这本书中来，但是有几点重要的信息是读者们应该知道的。

Why Android

根据 2012 年 11 月的数据，Android 占据全球智能手机操作系统市场 76%的份额，而中国市场占有率为 90%。Android 已经是而且将来也会继续稳坐智能手机头把交椅的宝座，因为不管什么厂商都可以生产 Android 手机，现在只要几百元就能买到一部看起来还不错的 Android 手机。

Android 的开发语言

Android 的 SDK 基于 Java 实现，但这不意味着必须用 Java 才可以编写 Android 程序。Google 公司在 2011 年初发布了 Android NDK(Native Development Kit)，这使得用 C 语言编写 Android 程序变得更加方便。当然，在没有对速度要求特别高的情况下，都是使用 Java 语言编写。不幸的是，本书没有涉及任何对速度要求特别高的应用。这也意味着，这是一本通俗的技术书。

Android 发展史

我知道如果我煞有介事地介绍 Android 从 1.0 到 4.2 各个版本的话，你们也不会看的，所以我打算介绍 Android 每个版本的特性（当然，我指的是原生 Android，不经过第三方改造的 Android 系统）。

1.0(图 1-1)：Android 的老大哥，首次搭载在 Google 太子 G1(HTC 代工)身上，这个版本拥有 Android 最基本的特性：可以下拉的通知栏，桌面插件（当然，打电话发短信上网什么的就不一一列举了）。

1.1：首创 OTA 升级功能（就是在不影响你手机的情况下升级系统，这下不用刷机了），但是因为各种问题，“胎死腹中”……

1.5：虚拟键盘（什么？Android 1.0 不支持虚拟键盘?)、第三方桌面控件（是的没错，

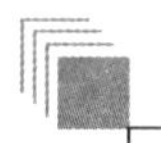

Android 1.0 只能支持手机自带的控件)。还有一点,从这个版本开始,Google 为每个 Android 版本起一个代号,代号首字母在字母表中的位置将能够反映这是 Android 的第几个版本,比如 CUPCAKE 的“C”意味着这是 Android 的第三个版本。所以 Android 1.5 也叫做“纸杯蛋糕”(Cupcake)。

1.6: 代号“甜甜圈”(Donut),亮点不多,增加了全局搜索功能。

2.0/2.0.1/2.1: 代号“松饼”(Eclair),增加动态壁纸(很好玩)、语音输入(不得不说 Google 的语音服务是比 iOS 早得多,只不过因为 Siri 是 Steve 的作品,所以才引起这么多关注),滑动解锁(没错,在这之前,一般是按 menu 键来解锁的)。

图 1-1 为 Android 1.x 解锁界面与 Android 2.0 的比较。

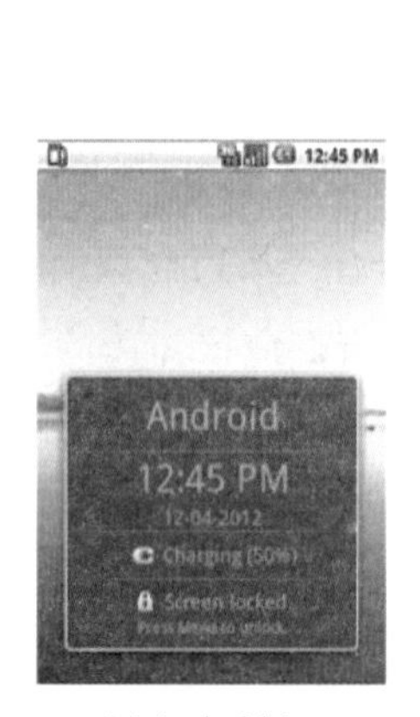

(a) Android 1.x

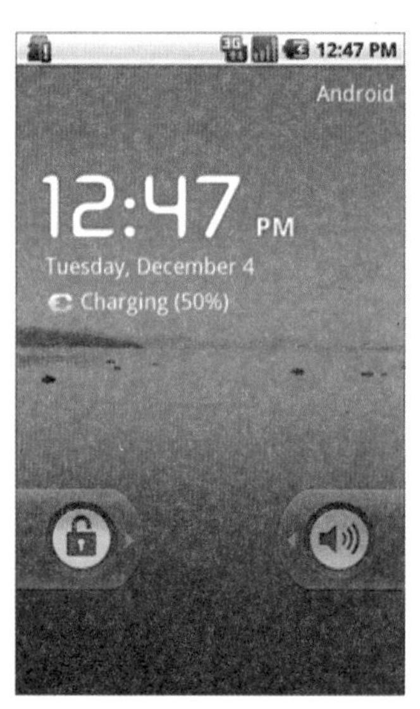

(b) Android 2.0

图 1-1 Android 1.x 解锁界面与 Android 2.0 的比较

2.2: 代号“冻酸奶”(Froyo),都是一些细节(这些细节有些厂商已经做到了,比如主屏幕数量,快捷方式,使用按钮取代抽屉来进入应用列表等),没有什么特别的东西。

2.3: 代号“姜饼”(Gingerbread),首次搭载在 Google 二皇子 Nexus S(三星代工; 同时,也是本书使用的手机,但是系统是 Android 4.0)身上,加强了复制粘贴功能,增加前置摄像头。

3. x(图 1-2): 代号“蜂巢”(Honeycomb),为平板而生,最终归并入 Android 大阵营。这个版本的特点有: 取消实体按键,增加 Action Bar、Fragment,超炫的多任务系统。特别的是,这个版本让我印象最深刻的是它自带的时钟控件。

4.0(图 1-3): 代号“冰激凌三明治”(Ice Cream Sandwich)。在经历了短暂的平板手机系统分家之后(Android 3.x 和其他),Google 把它们统一了起来,于是有了冰激凌三明治。搭载于 Google 三皇子 Galaxy Nexus(三星代工)身上,这个版本继承了蜂巢的特性,主题颜色由白变黑,使用全新 UI(至于有多全新不在这一一赘述),使用全新 Roboto 字体……最后一点,会有一个 Google 搜索框会如影随形地跟着你的屏幕。

4.1/4.2(图 1-4): 代号“果冻豆”(Jelly Bean),推出 Google Now,以此证明在搜索方面 iOS 不是其对手。

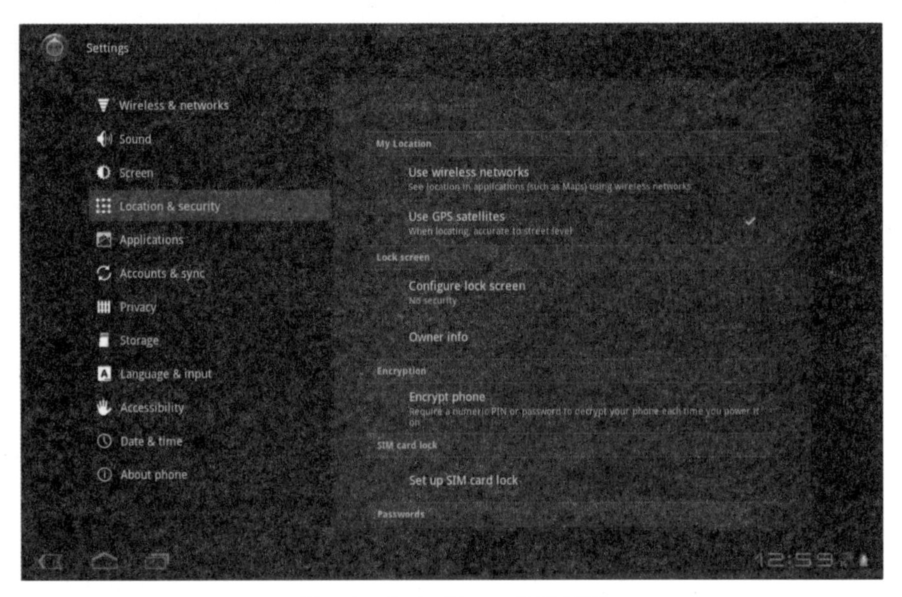

图 1-2 Android 3. x 设置界面

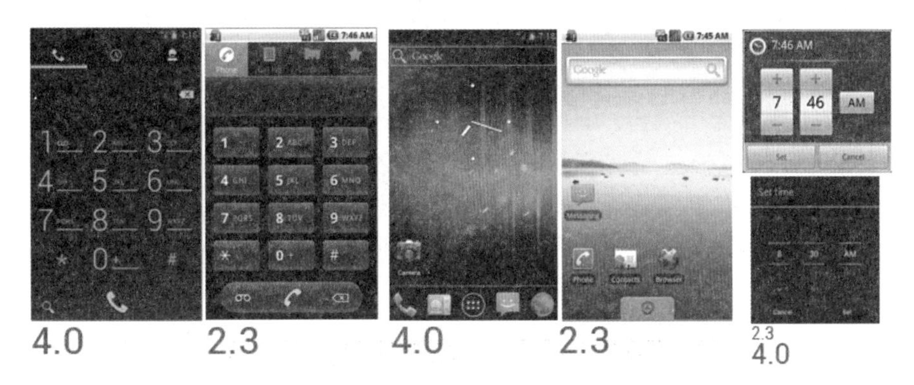

图 1-3 Android 4. 0 与 Android 2. 3 界面对比

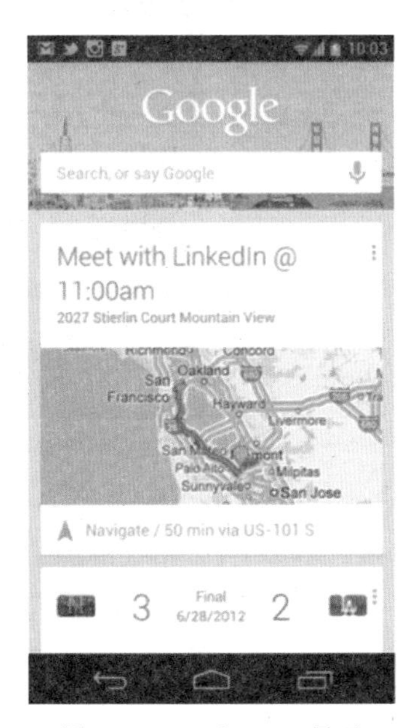

图 1-4 Google Now 界面

4.3(图 1-5)：代号依然是 Jelly Bean，于 2013 年 7 月 25 日推出，系统做出了多个细节改进，最大的变化是可以支持多用户。

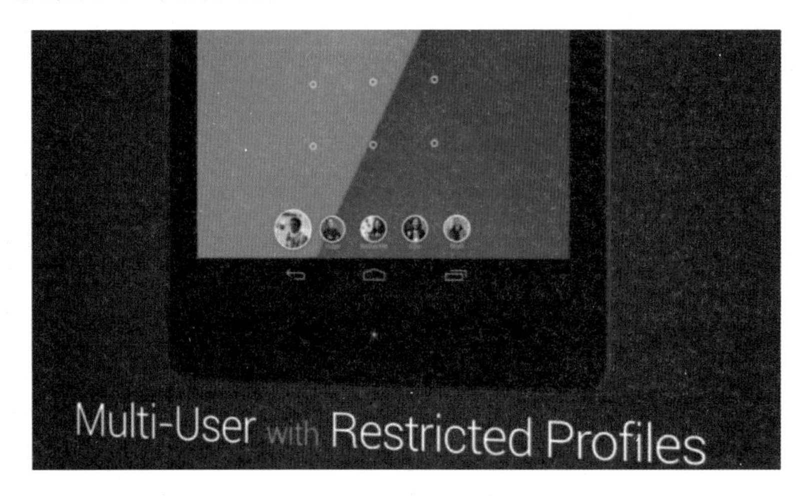

图 1-5　Android 4.3 支持多用户

我能用 Android 来做什么

很多很多，几乎是为所欲为。我个人认为界面无图标化会是未来的发展趋势。我们最需要的那几个功能可以通过手势来切换。碰巧的是，就在笔者写这一章的今天(2013 年 1 月 3 日)，Ubuntu 发布了移动版系统(图 1-6(a))，该系统的应用程序启动栏的操作与我在之前书中跟大家分享的“快速启动”(图 1-6(b))应用的操作有那么一点点神似。Android 可以让你做一些你觉得很酷的东西，让你捣鼓一些全新的体验，全新的设计。

(a) Ubuntu移动版　　(b) 快速启动

图 1-6　Ubuntu 移动版与快速启动

两年前我替别人写了一个课表的程序，只是因为在网上找不到一款专用的、好用的课表记录应用。后来这块细分领域有超级课程表和课程格子(图 1-7)，其中前者已经进行了千万级的 A 轮融资。

再后来，我受 Paul Graham 的启发，打算做一个让用户自己用图形界面“编程”，创建简单的应用。做到一半的时候发现 ITTT 火了，然后 Moto 的 Razr 手机自带了一个叫做“智能操作”的应用(图 1-8)。没错，这就是我想要做的东西。

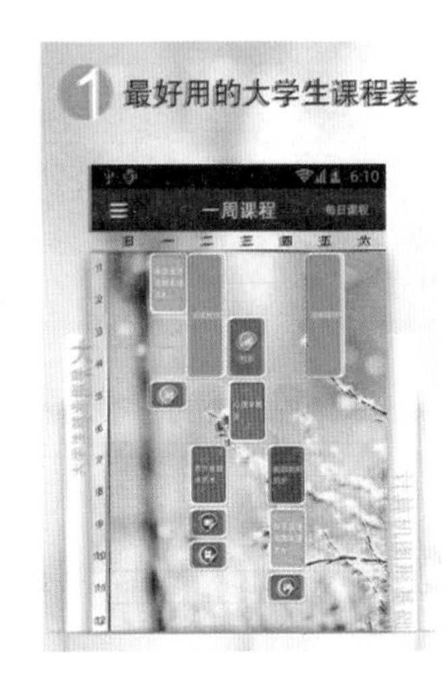

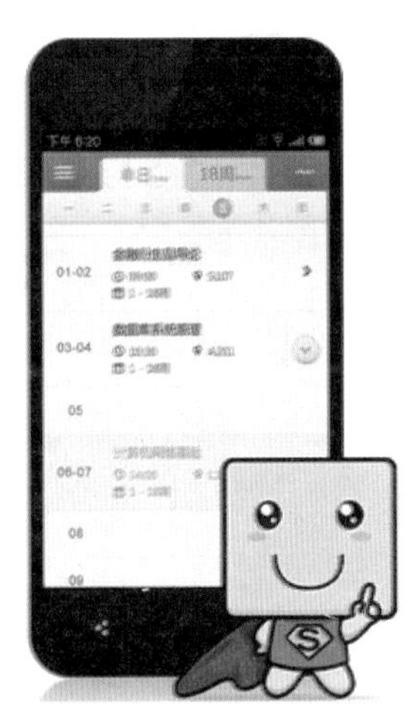

图 1-7　超级课程表和课程格子

图 1-8　Razr 手机的“智能操作”应用

想法总是会有的。这世界上有这么多人在发呆在思考，总会想出一样的东西。不同的是，那些成功的人一定会马上去做。创意本质上是不受法律保护的，你想得出来，我也想得出来。更何况，成功的往往是使用轮子的人，而不是发明轮子的人。

Android 就是轮子，同样 iOS 也是轮子，WP、WebOS 也是。如果你操作系统课的成绩不够好也没关系，你是使用轮子的那个人，而不是发明轮子的那个人。

Android 特性

浮窗

我记得最早在手机上看到浮窗这个东西的是当时雄霸一方的塞班。这让我印象很深刻，特别是使用某即时聊天工具的时候(你懂的)。Android 给开发者开放了这个 API(更确切地说应该是，添加系统窗口的 API，而不仅仅是浮窗)。有了这个接口你可以做很多有意思的东西，比如说让系统进入“宕机模式”等等。

替换主屏

这一点是 Android 中最酷的。想想看，你可以通过简单的代码替换原有的主屏，让你的手机变得独一无二。

替换解锁界面

既然可以替换主屏，那么替换掉解锁界面也是没问题的。

窗口小部件

Android 系统最直接表现的特点。窗口小部件是一个用户体验很好的东西，很方便实用。

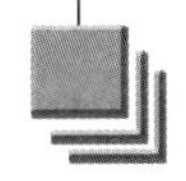

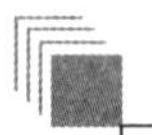

通知中心

曾经是，在被 iOS 等众多系统模仿了之后也不算是了。

本书特色

- 讲代码而不是贴代码。本书配套资料中有所有程序的代码源文件，并且是注释过的，我只会在书中介绍程序中的重要代码，图文并茂，尽量保证读者能看懂。
- 好看。诚然，我不是设计师，本书中的程序的界面风格不都是独创的。即便有些东西是我自己想的，但是谁又能保证，这些风格不是受到某些特定的影响在你脑海里留下来的？就算有人可以保证这点，又有谁能保证这个想法我是第一个想到的？“好的艺术家复制，伟大的艺术家偷窃”，如果你看过《硅谷传奇》，你会熟悉这句话的。我希望就算你没有把这本书看完，也能够非常优雅地摆放在你的书柜里。
- 关注移动互联网动态。本书中不只会介绍技术，更会介绍移动互联网发展动态，就像之前提到的 Ubuntu 移动版（当然，在我写这本书的时候是“最新动态”，等到出版的时候就是旧新闻了）。
- 关注用户体验。本书会引用交互设计中经典的案例、书籍，来说明为什么要这么做。

第二节　Android 环境搭建

这一节中，我们会一步一步地把环境搭建起来。Android 的开发环境搭建起来很方便，读者只需跟着我的步骤即可。

下载和安装 JDK

我们可以在 Oracle 的网站中下载到 JDK 进行 Java 的开发，登录网站 http://www.oracle.com/technetwork/java/javase/downloads/index.html，选择 JDK 进行下载，当前的最新版本是 Java 7。

在接下来的页面中选择要下载的版本，注意勾选 Accept License Agreement，并选择自己适合的版本，如图 1-9 所示。

下载完成后进行安装，如图 1-10 所示。

下载和安装 Eclipse IDE 以及 Android SDK

Android 提供了自带 Android SDK 以及 ADT 的 Eclipse，登录网站 http://developer.android.com/index.html 进行下载，选择 Get the SDK，如图 1-11 所示。

选择右边的大按钮进行下载，如图 1-12 所示。

下载完成之后，解压并打开根目录/eclipse/eclipse.exe。第一次打开的时候 Eclipse 会提醒你设置工作区间，这个工作区间会保存我们的配置信息、也是我们建立的工程所保存的路径。如果你不希望下次打开的时候再出现，可以勾选 Use this as the default and do not ask again，如图 1-13 所示。

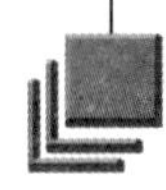

Java SE Development Kit 7u25

You must accept the Oracle Binary Code License Agreement for Java SE to download this software.

Accept License Agreement　Decline License Agreement

Product / File Description	File Size	Download
Linux ARM v6/v7 Soft Float ABI	65.12 MB	jdk-7u25-linux-arm-sfp.tar.gz
Linux x86	80.38 MB	jdk-7u25-linux-i586.rpm
Linux x86	93.12 MB	jdk-7u25-linux-i586.tar.gz
Linux x64	81.46 MB	jdk-7u25-linux-x64.rpm
Linux x64	91.85 MB	jdk-7u25-linux-x64.tar.gz
Mac OS X x64	144.43 MB	jdk-7u25-macosx-x64.dmg
Solaris x86 (SVR4 package)	136.02 MB	jdk-7u25-solaris-i586.tar.Z
Solaris x86	92.22 MB	jdk-7u25-solaris-i586.tar.gz
Solaris x64 (SVR4 package)	22.77 MB	jdk-7u25-solaris-x64.tar.Z
Solaris x64	15.09 MB	jdk-7u25-solaris-x64.tar.gz
Solaris SPARC (SVR4 package)	136.16 MB	jdk-7u25-solaris-sparc.tar.Z
Solaris SPARC	95.5 MB	jdk-7u25-solaris-sparc.tar.gz
Solaris SPARC 64-bit (SVR4 package)	23.05 MB	jdk-7u25-solaris-sparcv9.tar.Z
Solaris SPARC 64-bit	17.67 MB	jdk-7u25-solaris-sparcv9.tar.gz
Windows x86	89.09 MB	jdk-7u25-windows-i586.exe
Windows x64	90.66 MB	jdk-7u25-windows-x64.exe

图 1-9　选择合适的版本进行下载

图 1-10　Java 安装界面

第一次进入 Eclipse 环境，它会显示 Welcome 界面。单击左上角的还原按钮进入工作区间，如图 1-14 所示。

进入工作区间后，我们可以看到 Eclipse 的 Java 视图，如图 1-15 所示。

关于 Eclipse 的简单介绍我们就先到这里，在接下来的章节中我还会详细介绍它的使用。

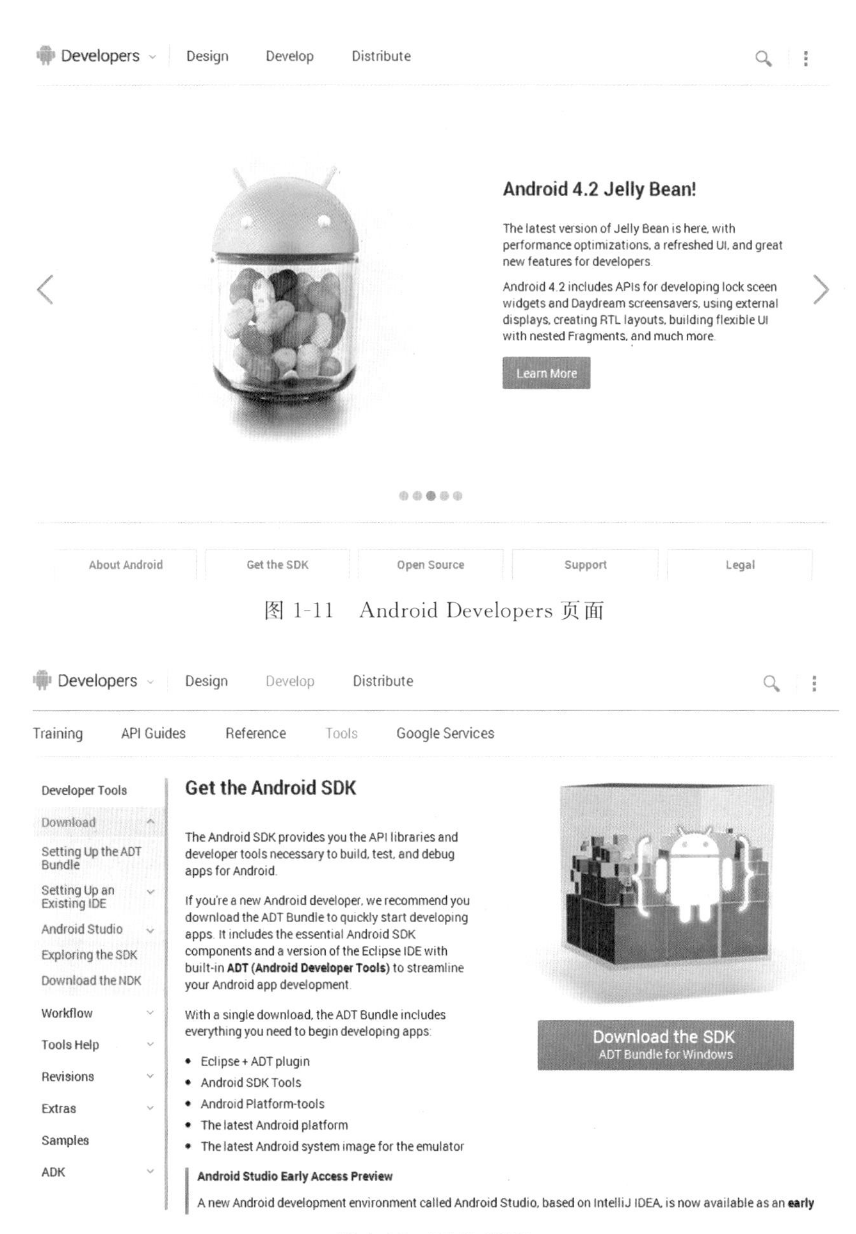

图 1-11 Android Developers 页面

图 1-12 下载 SDK

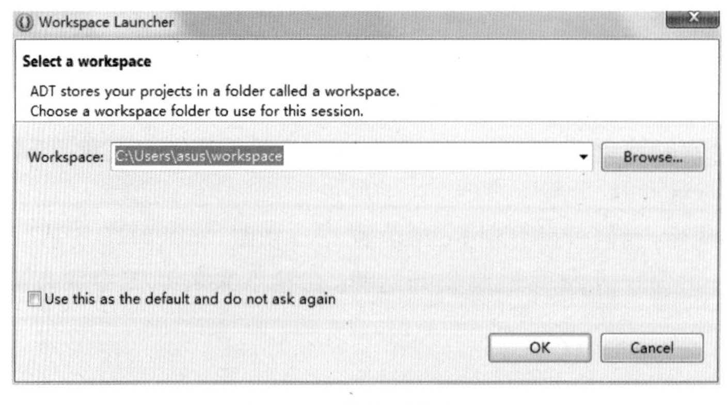

图 1-13 选择工作空间

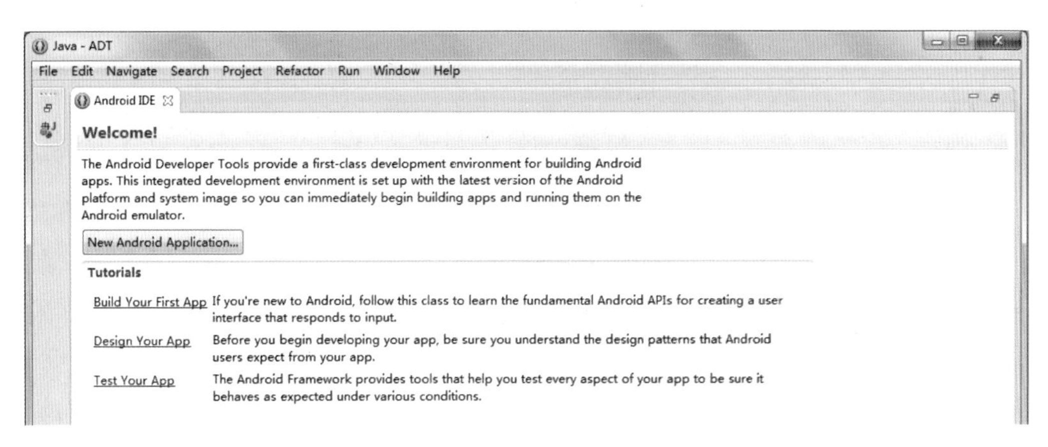

图 1-14 Eclipse 的 Welcome 界面

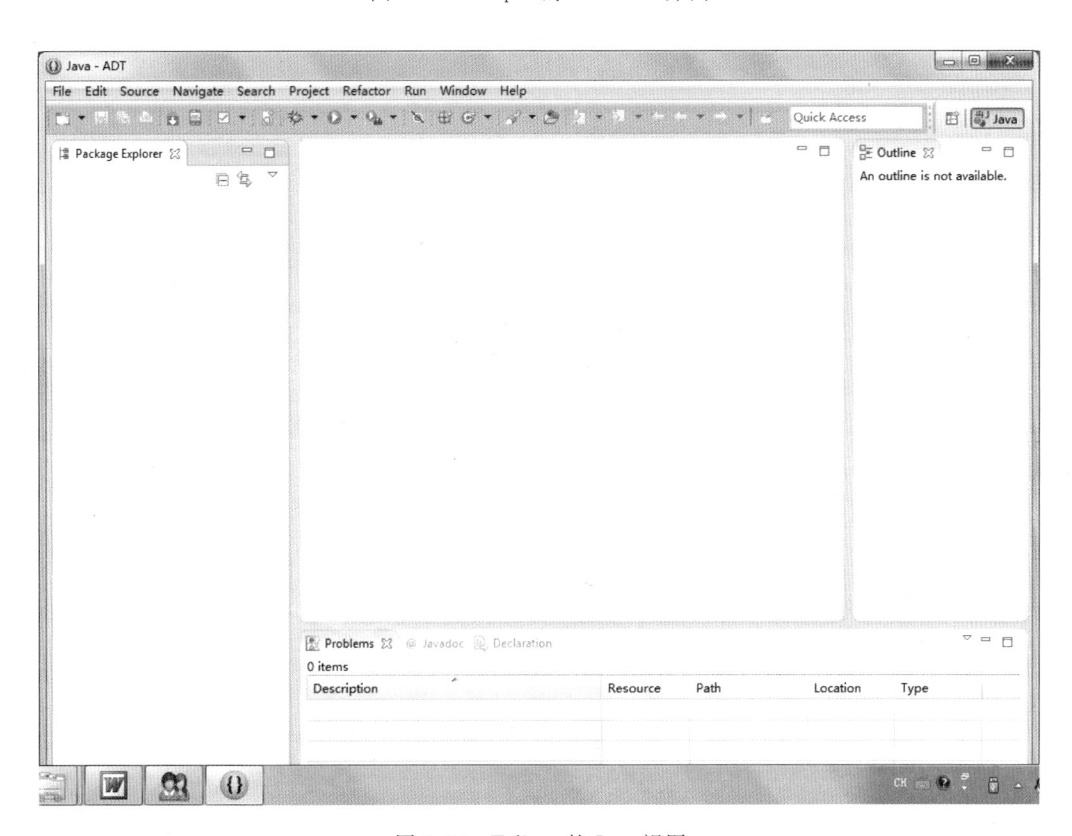

图 1-15 Eclipse 的 Java 视图

更新 Android SDK

从 Android 官网下载的 SDK bundle 包只包含了最新的 SDK 开发包，我们需要使用低版本的 SDK 进行开发，因此我们需要下载低版本的 SDK。如图 1-16 所示，选择 Window→Android SDK Manager。

勾选需要的 Packages，单击 Install packages，如图 1-17 所示。

选择 Accept License，单击 Install 进行安装，如图 1-18 所示。

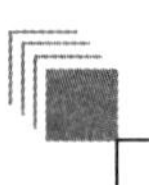

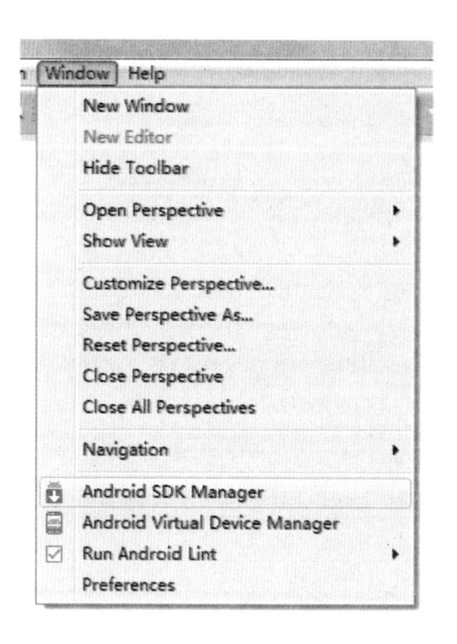

图 1-16　选择 Android SDK Manager

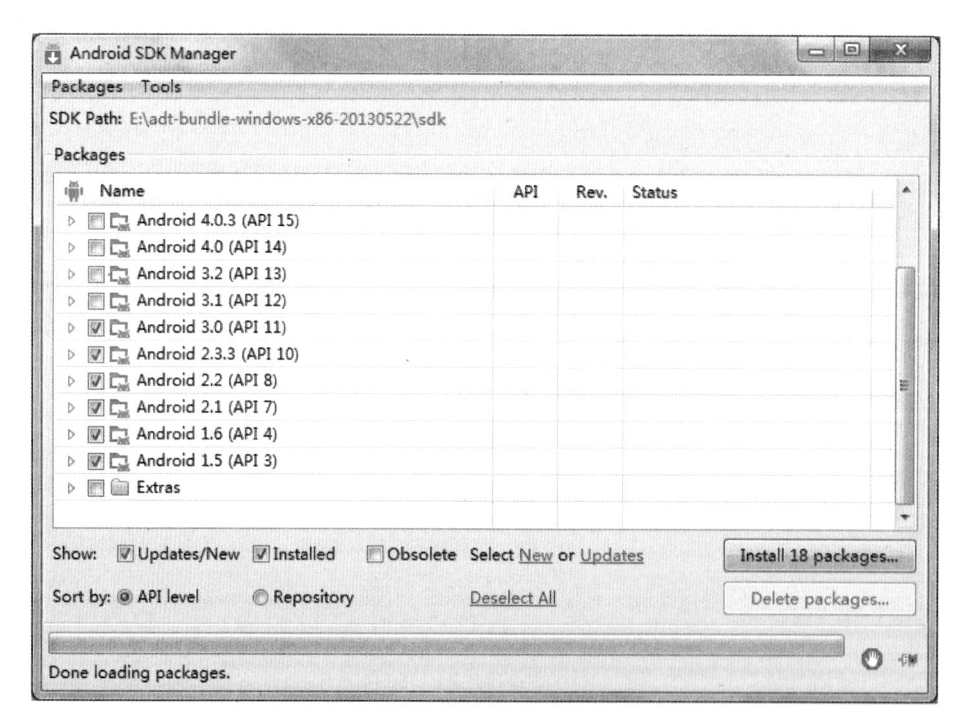

图 1-17　安装需要的包

安装完成之后，Android SDK Manager 会自动搜索可供下载的 SDK 版本，如图 1-19 所示。

这时候开始下载我们需要的 SDK，并在界面上动态显示下载过程，如图 1-20 所示。

当看到 Done loading packages 时，就说明 SDK 已经完成了下载，如图 1-21 所示。

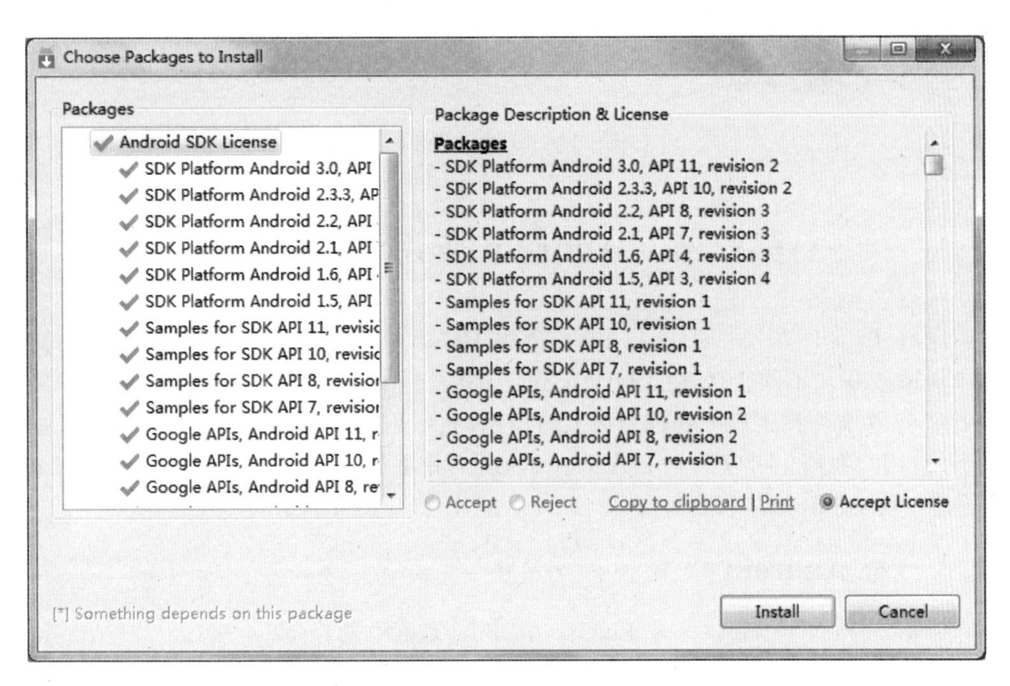

图 1-18　安装 Android SDK

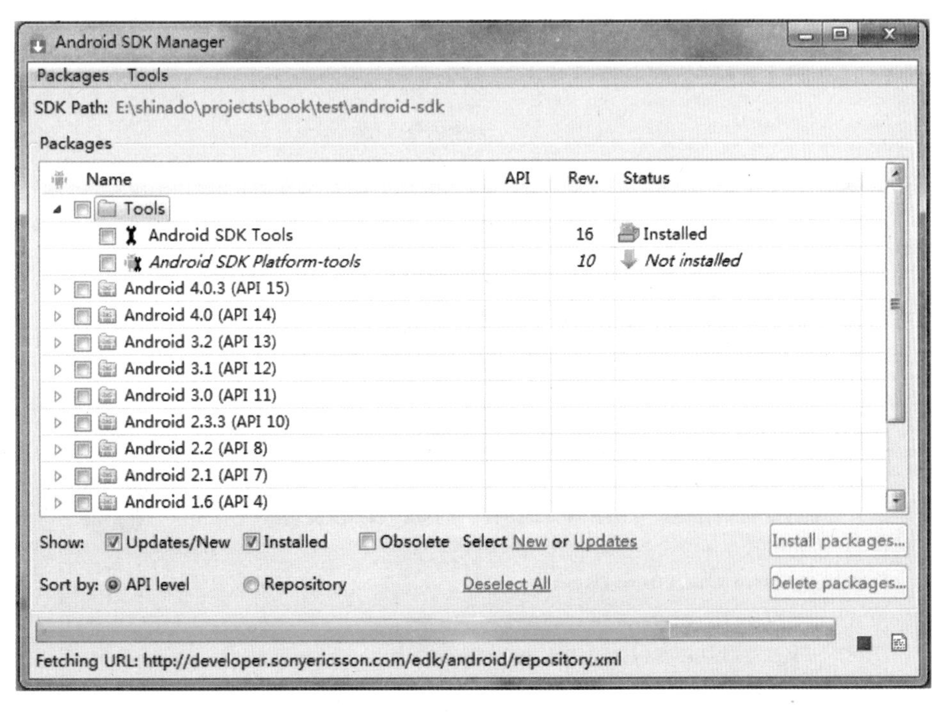

图 1-19　可供下载的 SDK 版本

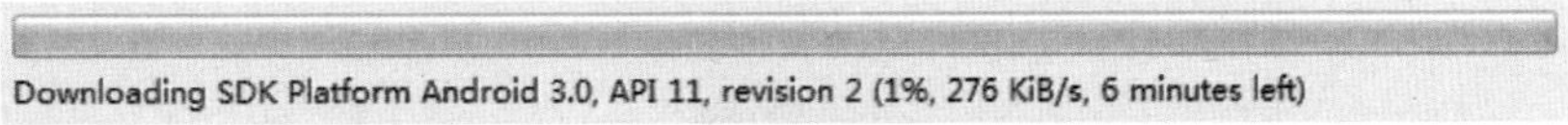

Downloading SDK Platform Android 3.0, API 11, revision 2 (1%, 276 KiB/s, 6 minutes left)

图 1-20　下载动态

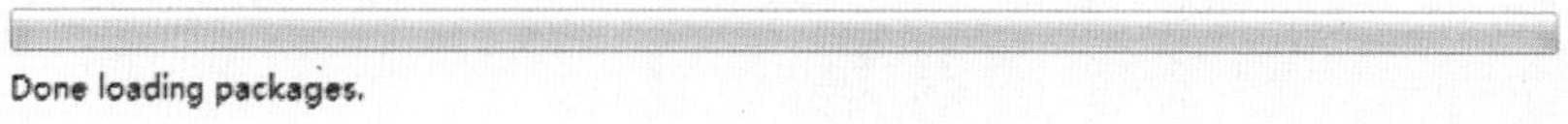

Done loading packages.

图 1-21　SDK 下载完成

第三节　Android 应用程序架构

搭建好开发环境后，就可以跟这个世界"say hello"了。当然，我指的是创建 HelloWorld 工程。不同于普通的 Java 工程，一个 Android 工程包含了除 Java 代码之外的许多文件，这些文件在工程中发挥着很大的作用，它们可以帮助我们更方便、更灵活地开发 Android 应用。我们也可以把一个 Android 工程看作框架。这一节我们就来讲 Android 工程的创建以及 Android 工程的框架。

创建第一个 Android 项目

第一步，选择创建工程：打开 Eclipse，在 File 菜单中选择 New→Android Application Project，如图 1-22 所示。在 Android 包中选择 Android Project，单击 Next 按钮。

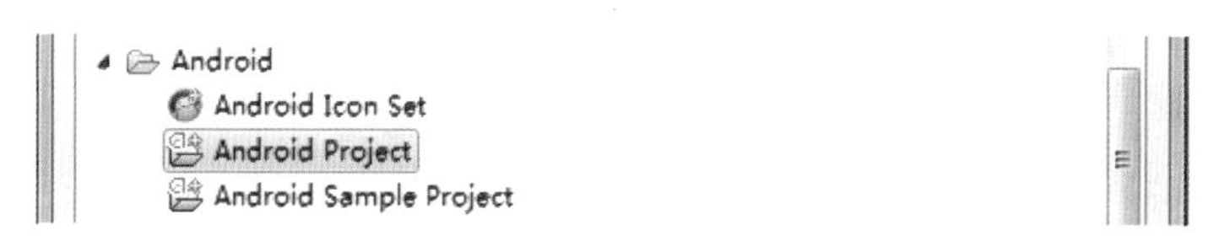

图 1-22　选择创建工程

知识扩充

如果看不到 Android Application Project 这个选项，在 File 菜单中选择 New→Other…，在 Android 包中选择 Android Project，单击 Next 按钮。

第一个页面(图 1-23)，填写应用名、工程名和包名，选择 Build SDK，也就是使用哪个版本的 SDK 开发。由于 Android 是向上兼容的，所以尽量用低版本进行开发。

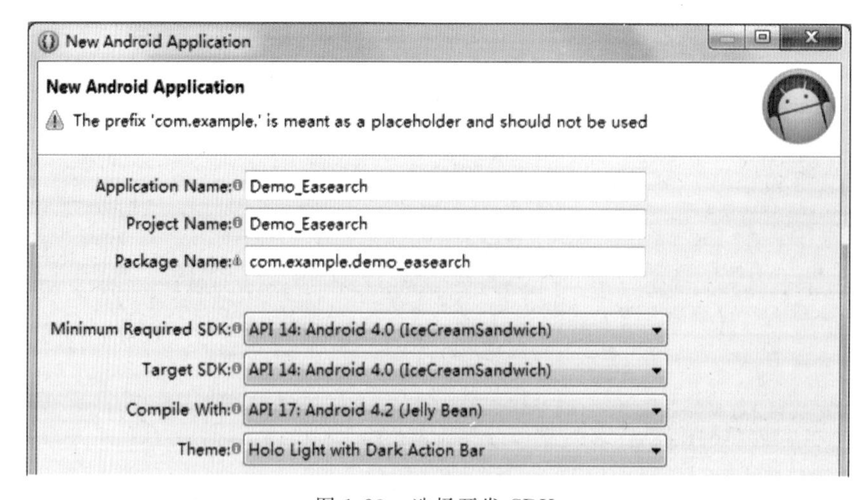

图 1-23　选择开发 SDK

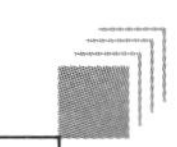

第二个页面(图 1-24)用来自定义图标,没多大用处,因为我们一般都会使用自己的图标。

图 1-24 选择图标

第三个页面(图 1-25)选择是否要创建 Activity,如果勾选的话还有第四步,没有的话就可以直接完成。由于我们使用的 SDK 版本为 7,不能支持 Fragment,所以只能选择 BlankActivity。

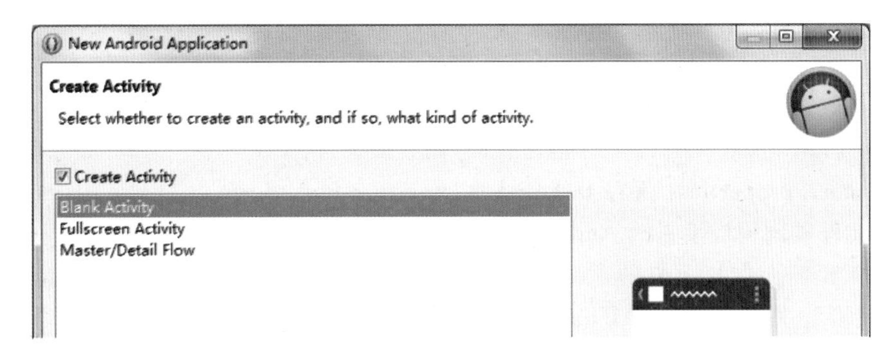

图 1-25 创建 Activity

第四个页面(图 1-26)设置 Activity 的名称、布局。由于我们使用的 SDK 版本不高,不能支持导航类型,所以也只能选择 None。

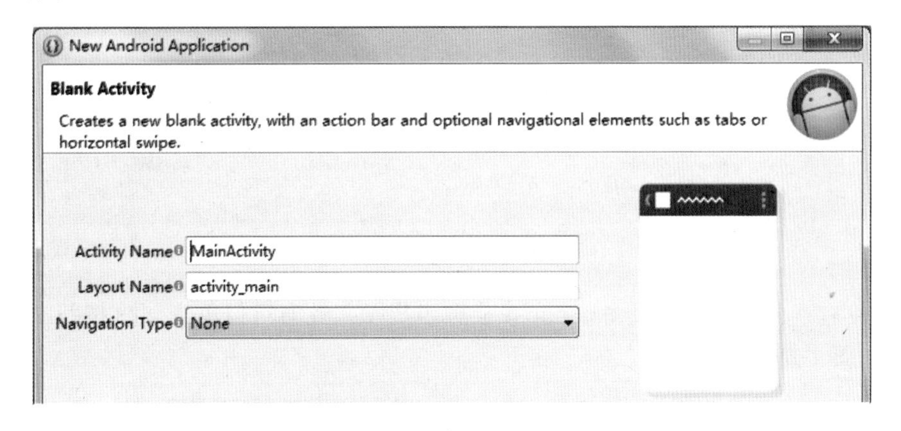

图 1-26 新建一个 Activity

知识扩充

不同版本的 ADT 创建 Android 程序的过程可能不同,但是大同小异。理论上来说,填好包名、应用名后点 Next 按钮即可。

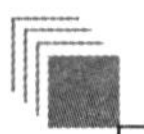

如图 1-27 所示，创建完成之后，我们可以看到一个 Android 工程默认包含了如下文件：

- src 文件夹：Java 源代码文件。
- gen 文件夹：存放 R. java 文件。
- R. java：封装静态变量，根据 drawable 文件、values 文件和 layout 文件动态改变。
- Android Dependencies：Android 库。
- assets 文件夹：存放不需要编译成二进制的文件。
- bin 文件夹：存放 apk、图片资源等二进制文件。
- res 文件夹：包含 drawable 文件夹、layout 文件夹和 values 文件夹。
- drawable 文件夹：存放图片文件(格式为 PNG)。
- layout 文件夹：存放布局文件(格式为 xml)。
- menu 文件夹：存放菜单布局文件(格式为 xml)。
- values 文件夹：存放参数文件。
- AndroidManifest. xml：Android 配置文件。
- proguard-project. txt：用于混淆代码，防止反编译。
- project. properties：保存 target 信息。

这些文件共同构成了一个 Android 的架构。这些文件的详细解释我们会在接下来的章节中介绍。但是一般来说，我们只需要关心 AndroidManifest. xml 文件和 res 文件夹，前者是 Android 的配置文件，后者是 Android 的资源文件。此外，R. java 文件是系统自动生成的，开发者无需手动修改。

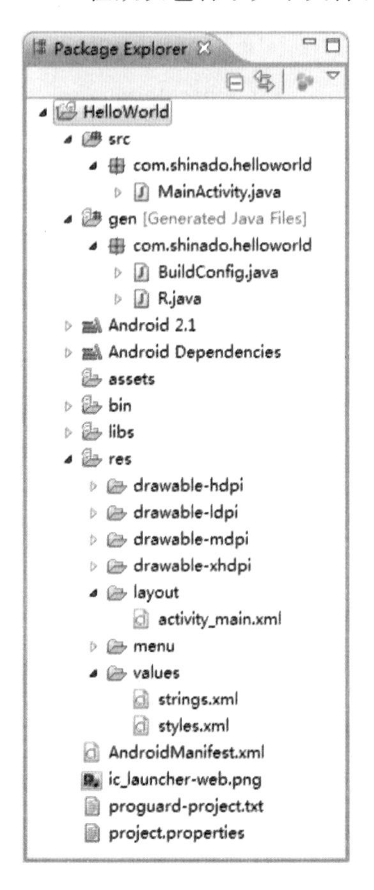

图 1-27　一个 Android 工程的框架

我们运行一下这个应用，在工程上右击，选择 Run as→Android Application，如图 1-28 所示。

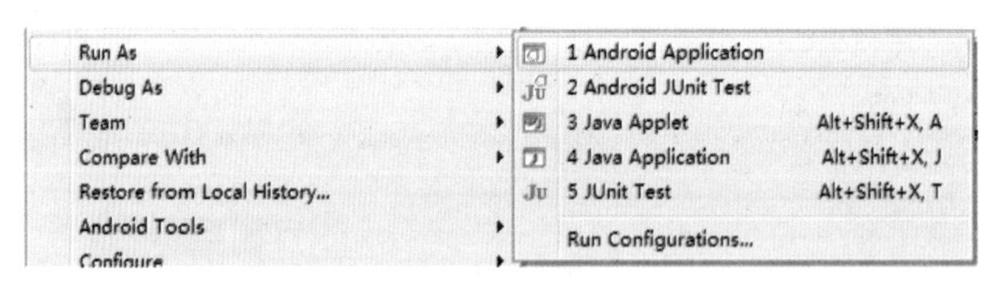

图 1-28　运行工程

如果没有连接 Android 设备，Eclipse 会弹出提示。此时可以单击 Yes 按钮来创建一个模拟器，如图 1-29 所示。

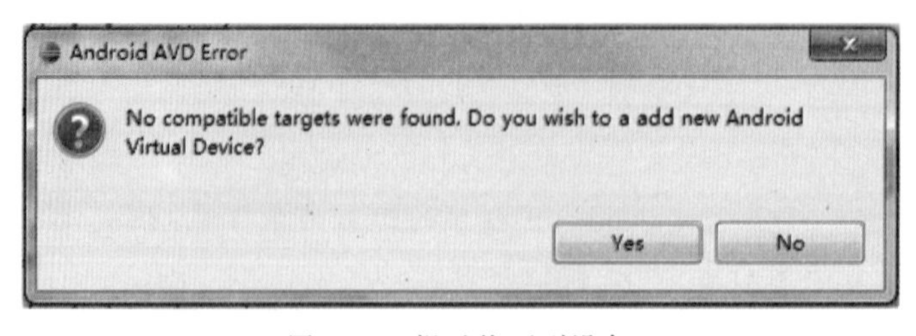

图 1-29　提示找不到设备

弹出模拟器管理窗口，单击 New 按钮进行创建，如图 1-30 所示。

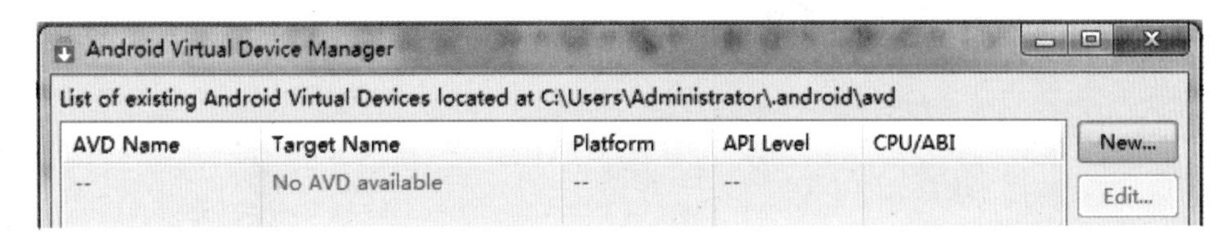

图 1-30　Android 模拟器管理

在创建 AVD 窗口中填写 Name、选择 Target(Android 2.1)、在 Size 一栏中填写“100”，单击 Create AVD，这样一来我们就创建了一个拥有 100MB 内存的 Android 2.1 的模拟器了，如图 1-31 所示。

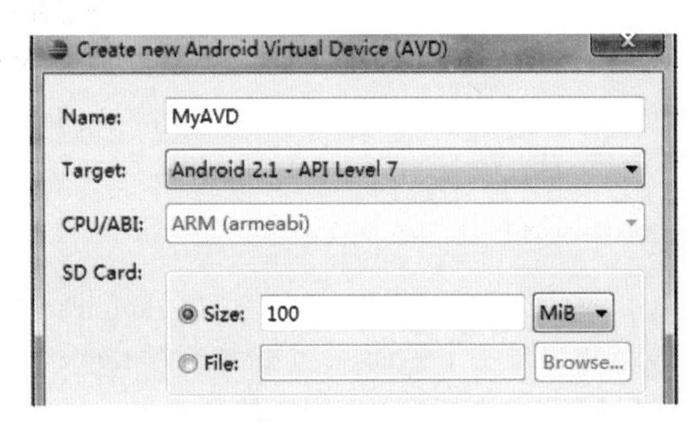

图 1-31　创建模拟器

创建完成后，返回到 Android Virtual Device Manager 窗口。此时，你创建的模拟器就会显示在列表中，如图 1-32 所示。

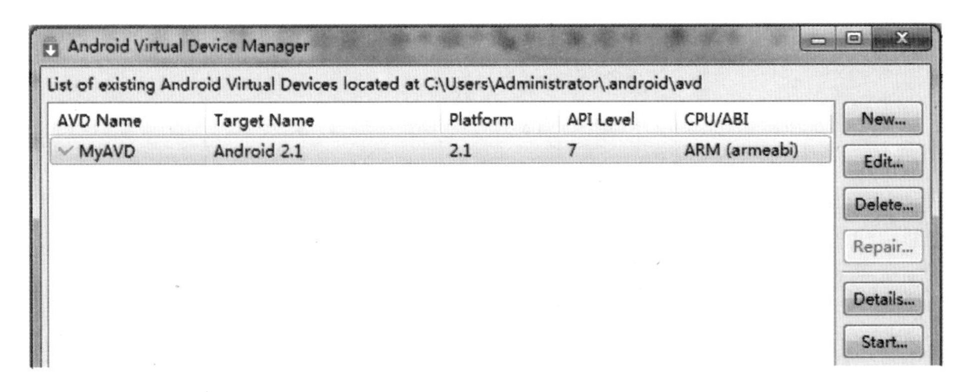

图 1-32　启动模拟器

单击 Start 按钮，弹出确认窗口，单击 Launch 按钮，启动模拟器并运行我们的应用，如图 1-33 所示。

在模拟器上就可以看到我们的应用，如图 1-34 所示。

值得注意的是，模拟器的启动较慢，所以我们在使用模拟器开发应用的时候不需要把模拟器关闭，避免重启模拟器的麻烦。

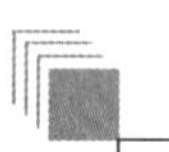

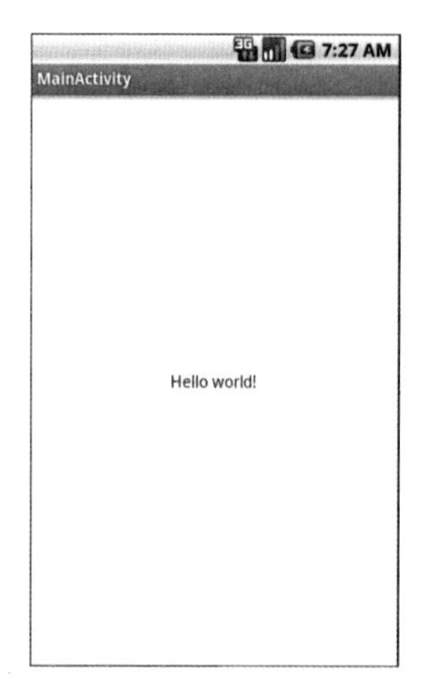

图 1-33　工程的运行结果

图 1-34　运行完成的应用

AndroidManifest. xml 文件

我们要先了解 activity 的概念：一个 Android 应用(application)，它可能包含多个 activity。而一个 activity 就相当于一个界面、一个窗口。Activity 类似于 Windows 中“窗口”的概念，用于与用户进行交互。Android 应用中的每一个 activity 都要继承于 Activity 这个类，并且在 AndroidManifest. xml 文件中标明出来。AndroidManifest. xml 文件描述了一个 Android 应用的应用名称、图标、包含多少个组件、权限等信息。

我们先来看创建的 HelloWorld 工程生成的 AndroidManifest. xml 文件：

```
<manifest xmlns:android = "http://schemas.android.com/apk/res/android"
    package = "com.shinado.helloworld"
    android:versionCode = "1"
    android:versionName = "1.0" >

    <uses - sdk
        android:minSdkVersion = "7"
        android:targetSdkVersion = "15"/>

    <application
        android:icon = "@drawable/ic_launcher"
        android:label = "@string/app_name"
        android:theme = "@style/AppTheme" >
        <activity
            android:name = ".MainActivity"
            android:label = "@string/title_activity_main" >
            <intent - filter>
                <action android:name = "android.intent.action.MAIN"/>

                <category android:name = "android.intent.category.LAUNCHER"/>
            </intent - filter>
```

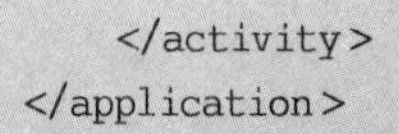

```
        </activity>
    </application>

</manifest>
```

最外层的节点为<manifest>，它的常用属性有如下几个。

- package：包名。
- android:versionCode：版本号。
- android:versionName：版本名称。

常用的第二层节点有如下几个。

- uses-sdk：描述 SDK 属性，包含：
    - ◆ android:minSdkVersion，需要最低的 SDK 版本。
    - ◆ android:targetSdkVersion，最适合的 SDK 版本，建议删去。
- uses-permissions：包含属性 android:name，用于描述权限。
- application：描述应用的组件、应用名等信息，包含属性：
    - ◆ android:icon：描述应用的图标。
    - ◆ android:label：描述应用名。
    - ◆ android:theme：描述应用主题。

第三层节点包含于 application，包括 activity、service、receiver 和 provider 等。

我们在这里先讨论 activity。activity 是一个程序中跟用户交互的组件，它必须继承于 Activity 这个类，这里只是声明了这个 activity 的位置。

```
<activity
    android:name=".MainActivity"
```

代码中包含了一个点“.”，这表示该类在声明的包下：

```
package="com.shinado.helloworld"
```

也就是说，它指向了包 com. shinado. helloworld 指向了 MainActivity 这个类。当然，如果你不幸在另外一个包里写了一个 Activity，那么就要在 android:name 节点中声明完整的类的地址。

- android:name：Activity 的位置。
- android:label：用于设置该 Activity 的标题。

第四层 intent-filter 是一个过滤器，用于过滤广播、动作等信息，它包含了第五层节点。

- ◆ action：包含属性 android:name，一般我们常见到的 android. intent. action. MAIN 表示这个 activity 是该应用的入口。
- ◆ category：包含属性 android:name，一般我们常见的 android. intent. category. LAUNCHER 表示应用会出现在程序的列表中。

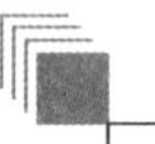

资源文件夹

我们可以看到 res 文件夹里包含了 7 个子文件夹，如图 1-35 所示。

drawable 文件夹

drawable 文件夹有如下 4 个。

- drawable-xhdpi：存放超高分辨率的图片，一般为平板。
- drawable-hdpi：存放高分辨率的图片，如 WVGA (480×800)，FWVGA(480×854)。
- drawable-mdpi：存放中等分辨率的图片，如 HVGA(320×480)。
- drawable-ldpi：存放低分辨率的图片，如 QVGA (240×320)。

res
drawable-hdpi
drawable-ldpi
drawable-mdpi
drawable-xhdpi
layout
menu
values

图 1-35　res 文件夹

这样的好处是系统可以通过手机的分辨率来使用不同分辨率下的图片，实现屏幕自适应。关于这一点，我们会在第 2 章中继续讲解。

细心的读者可能会发现在 AndroidManifest. xml 文件中出现过“@drawable/ic_launcher”这样的字符串，它表示引用了 drawable 文件夹下的 ic_launcher 资源。

layout 文件夹

再来看 layout 文件夹，这个文件夹存放应用的布局文件，main. xml 文件代码如下：

```
<RelativeLayout
    xmlns:android="http://schemas.android.com/apk/res/android"
    xmlns:tools="http://schemas.android.com/tools"
    android:layout_width="fill_parent"
    android:layout_height="fill_parent" >

    <TextView
        android:layout_width="wrap_content"
        android:layout_height="wrap_content"
        android:layout_centerHorizontal="true"
        android:layout_centerVertical="true"
        android:text="@string/hello_world"
        tools:context=".MainActivity"/>

</RelativeLayout>
```

这个布局文件表示在一个相对布局中有一个居中的 TextView(也就是普通文本)，它显示的文本是 string 资源中的 hello_world 字符串(注意，不是字符串“hello_world”)。

在布局文件中一般有这样几个节点：

- android:id：以“@+id/”开头，声明控件的 id，以便于在 Java 代码中通过 findViewById 方法获得。
- android:layout_width：控件的宽度，可选参数：
  - fill_parent：填充父容器。
  - wrap_content：填充子容器。
  - 数值，以 px/dp 等单位结束。

- android:layout_height：控件的长度。可选参数与 android:layout_width 相同。
- android:gravity：控件子容器的重心。
- android:layout_gravity：控件相对于父容器的重心。

values 文件夹

values 文件夹下的 string. xml 代码如下：

```
<resources>

    <string name="app_name">HelloWorld</string>
    <string name="hello_world">Hello world!</string>
    <string name="menu_settings">Settings</string>
    <string name="title_activity_main">MainActivity</string>

</resources>
```

我们可以看到名为“hello_world”的 string 节点的值为“Hello World!”。这也就是说最终在程序中显示的文本为“Hello World!”,如图 1-36 所示。

最后一点需要注意的是,资源文件只能以小写字母和下划线作首字母,随后的名字中只能出现[a-z0-9_.]这些字符,否则就会造成错误。

```
[2012-12-07 11:19:48 - ShLauncher]res\drawabie\ABC.png: Invalid file name: must contain
only[a-z0-9-.]
```

除了 string 文件之外,还会有 color 和 style 文件。前者保存颜色代码,后者保存样式代码。layout 文件中经常会引用这几个常用文件：

- @string/：引用 strings. xml 中的文字。
- @color/：引用 color. xml 中的颜色代码。
- @drawable/：引用 drawable 中的图片或者 xml 文件。
- @style/：引用 styles. xml 中的样式。

menu 文件夹

menu 文件夹保存菜单的设置,当我们按下 menu 键的时候,就会显示该菜单,如图 1-36 所示。

当然,你需要重写 Activity 的方法才能实现。这点后面会详细讲。

图 1-36　menu 菜单

```
<menu xmlns:android="http://schemas.android.com/apk/res/android">
    <item android:id="@+id/menu_settings"
        android:title="@string/menu_settings"
        android:orderInCategory="100"/>
</menu>
```

R. java

先来看 R. java 的代码：

```
public final class R {
    public static final class attr {
    }
    public static final class drawable {
        public static final int ic_action_search = 0x7f020000;
        public static final int ic_launcher = 0x7f020001;
    }
    public static final class id {
        public static final int menu_settings = 0x7f070000;
    }
    public static final class layout {
        public static final int activity_main = 0x7f030000;
    }
    public static final class menu {
        public static final int activity_main = 0x7f060000;
    }
    public static final class string {
        public static final int app_name = 0x7f040000;
        public static final int hello_world = 0x7f040001;
        public static final int menu_settings = 0x7f040002;
        public static final int title_activity_main = 0x7f040003;
    }
    public static final class style {
        public static final int AppTheme = 0x7f050000;
    }
}
```

R. java 这个文件是系统自动生成的，当你在 res 文件中添加图片、布局文件、values 字段、布局中的 id 等时，R. java 就会自动生成静态的 int 变量。

我们在 main. xml 文件中的 TextView 节点中加入 id：

```
<TextView
        android:id = "@+ id/test"
```

保存之后，R. java 会自动在内部类 id 中增加一个静态 int 型变量 test（也就是我们定义的 id）：

```
public static final int test = 0x7f070000;
```

程序的实现原理

上一小节中，我们把 main. xml 中的 TextView 增加了一个 id。此时，我们修改一下 src 文件夹下的文件 MainActivity. java：

```
@Override
public void onCreate(Bundle savedInstanceState) {

    super.onCreate(savedInstanceState);
    setContentView(R.layout.activity_main);

    //添加的内容
    TextView tv = (TextView) findViewById(R.id.test);
    tv.setTextSize(30);
    tv.setText("text changed");
}
```

setContentView 方法是 Activity 中最基本的方法之一，它定义了 Activity 的界面显示内容，而这个界面的内容就是写在 layout 文件夹的文件中。在本例中，setContentView(R.layout.main)就表示设置 Activity 的界面布局为 layout 文件夹下的 main.xml 文件所定义的布局。

运行一下，程序的截图如图 1-37 所示。

我们可以看到，程序显示的文本变成了"text changed"，字体也变大了。这是因为我们通过反射获得 layout 文件中定义的 TextView 组件（也就是 findViewById(R.id.test)这个方法），并修改这个组件的文本和字体大小。

在运行 Android 程序时，系统先查看 AndroidManifest.xml 文件，看程序的入口是哪个 Activity；找到那个 Activity 之后看这个 Activity 设置的是哪个 layout 文件（setContentView 方法），然后加载该布局文件，显示在界面上。AndroidManifest.xml、Activity、layout 文件、values 文件和 R.java 文件的关系如图 1-38 所示。

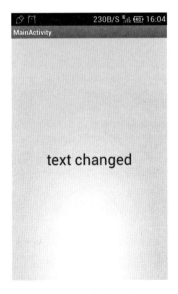

图 1-37　修改文字

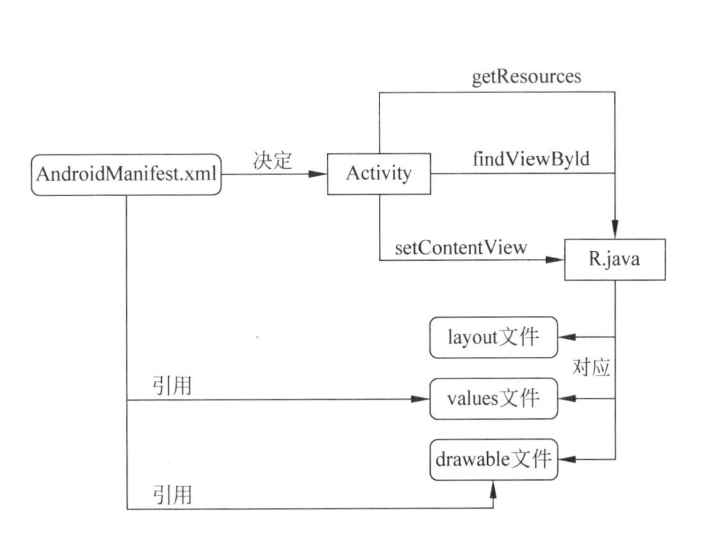

图 1-38　关系图

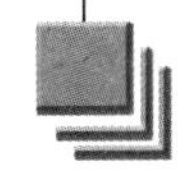

第四节　使用 ADT

LogCat

一般来说，在 Java 中我们使用 System.out.println 方法来输出某个值。如果用户在代码中使用了 System.out.println 方法，那么不会在控制台里看到输出。因为 Android 把输出的结果交给了 LogCat 工具。

首先，我们要调出 LogCat 窗口，在 Eclipse 中的 Window 菜单中选择 Show View→Other，在 Show View 窗口中的 Android 文件夹中选中 LogCat（你会看到这里有两个 LogCat，其中一个声明 deprecated，也就是不建议使用），单击 OK 按钮，如图 1-39 所示。

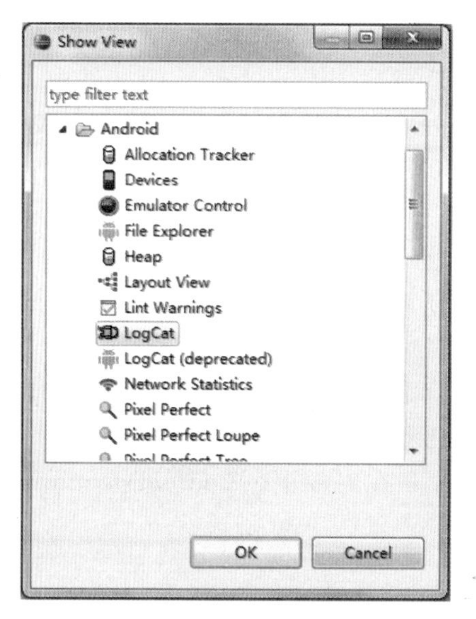

图 1-39　显示 LogCat

此时，Eclipse 中的输出栏就会多一个 LogCat 窗口，如图 1-40 所示。

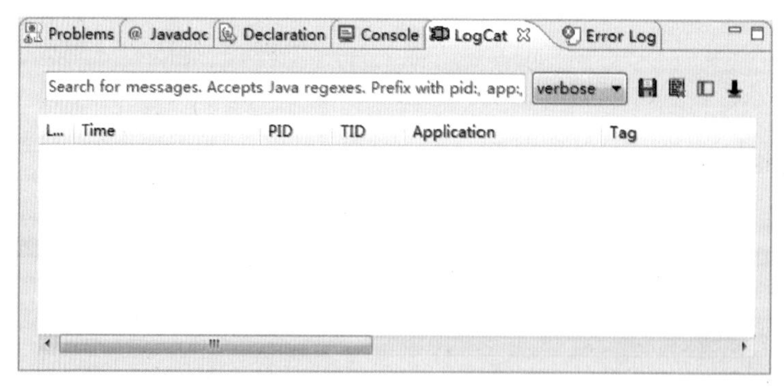

图 1-40　LogCat 窗口

我们可以通过右上方的圆形按钮来切换 Log 的内容。

Log 常用的方法有以下 5 个。

- Log.v(String tag,String msg)：对应圆形按钮 V，表示啰嗦。

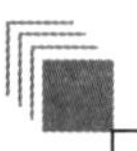
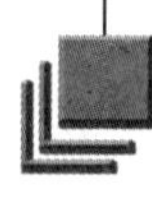

- Log. d(String tag,String msg)：对应圆形按钮 D,表示 debug。
- Log. i(String tag,String msg)：对应圆形按钮 I,表示提示信息。
- Log. w(String tag,String msg)：对应圆形按钮 W,表示警告。
- Log. e(String tag,String msg)：对应圆形按钮 E,表示错误。

我们来试一下 Log 的使用,在 HelloWorldActivity. java 中的 OnCreate 方法中添加代码：

```
Log.v("啰嗦","verbose");
Log.d("调试","debug");
Log.i("信息","information");
Log.w("警告","warnning");
Log.e("错误","error");
```

查看一下 LogCat 会看到 Eclipse 下方的输出信息,看起来色彩斑斓,如图 1-41 所示。

Declaration | Console | LogCat (deprecated) | LogCat

Search for messages. Accepts Java regexes. Prefix with pid:, app:, tag: or text: to limit scope.

L...	Time	PID	Application	Tag	Text
V	01-02 12:18:16.481	320	com.shinado.hello...	啰嗦	verbose
D	01-02 12:18:16.491	320	com.shinado.hello...	调试	debug
I	01-02 12:18:16.510	320	com.shinado.hello...	信息	information
W	01-02 12:18:16.510	320	com.shinado.hello...	警告	warnning
E	01-02 12:18:16.510	320	com.shinado.hello...	错误	error

图 1-41　LogCat 输出信息

DDMS(Dalvik Debug Monitor Server)

DDMS 是个非常好用的东西,它能给我们提供设备截屏,查看文件(删了 360 手机助手吧)、查看进程、线程以及堆信息、查看广播状态信息、模拟电话呼叫、接收短信、模拟地理坐标等功能。开发者可以通过 Eclipse 的右上角的添加按钮 Java 来添加 DDMS 视图,如图 1-42 所示。

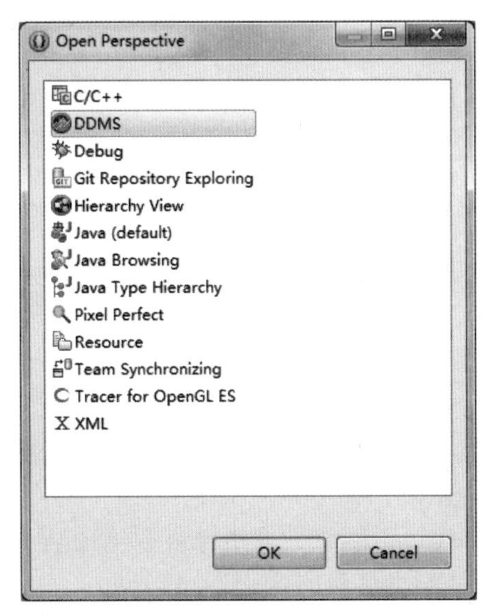

图 1-42　添加 DDMS 视图

窗口的左侧是设备以及设备的进程信息，右侧是各个功能的视图，如图 1-43 所示。

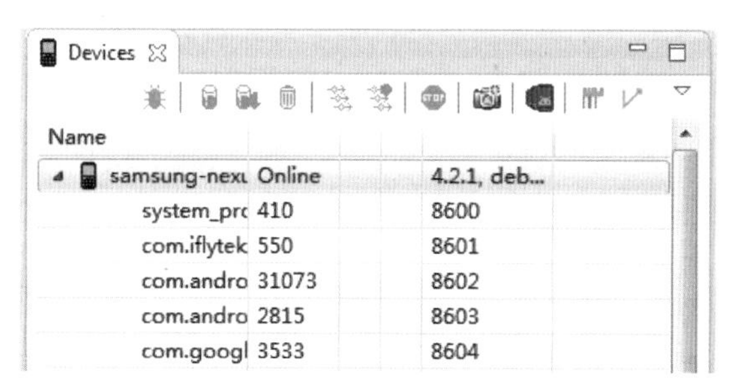

图 1-43　设备进程信息

截图

在左边的窗口中选中设备，单击照相的小按钮，出现手机屏幕的实时图像（当然手机要连上电脑），如图 1-44 所示。

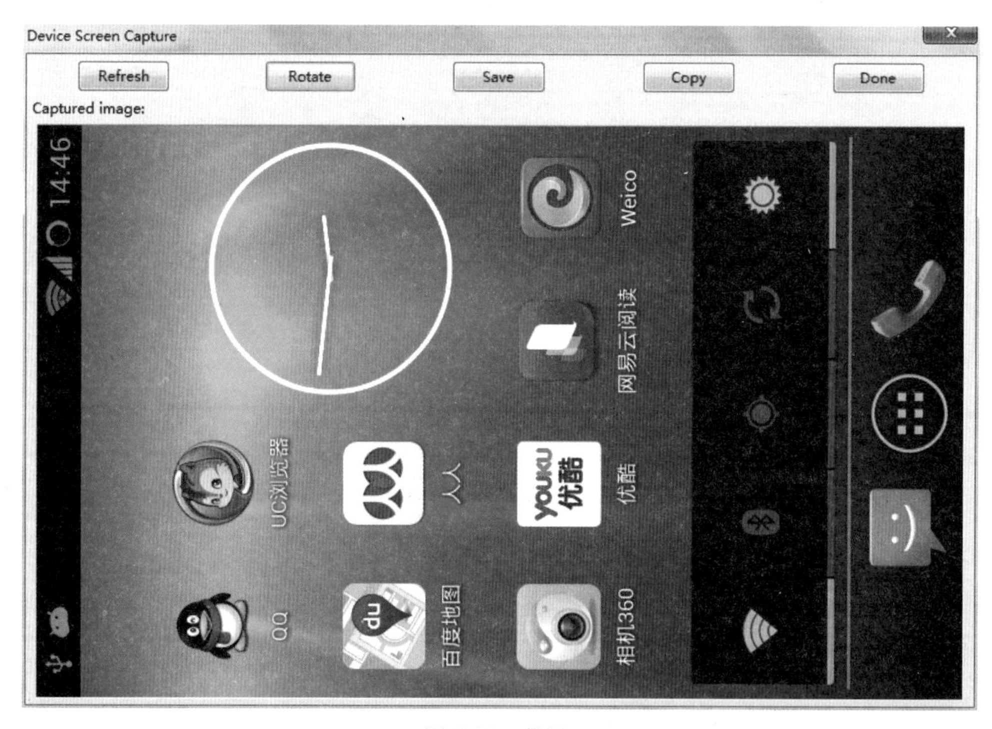

图 1-44　截屏

查看 UI 布局

最新版的 ADT 中增加了这个强大的功能，选中设备后单击手机的小按钮，可以查看界面的 UI 布局，如图 1-45 所示。

查看线程

在窗口左侧的进程栏中，选择要查看的进程，单击 Upgrade Threads 按钮，这个进程就会出现一个线程小图标，如图 1-46 所示。

在右侧的窗口中选择 Threads 栏，就会显示该进程下的所有线程，双击某个线程，线程框的下方会显示这个线程是从哪里启动的，如图 1-47 所示。

图 1-45　查看 UI 布局

图 1-46　更新进程信息

Threads | Heap | Allocation Tracker | Network Statistics | Emulator Contr

ID	Tid	Status	utime	stime	Name
1	14464	native	474	46	main
*2	14467	vmwait	14	0	GC
*3	14469	vmwait	0	0	Signal Catcher
*4	14470	running	21	6	JDWP
*5	14471	vmwait	8	5	Compiler
*6	14472	wait	0	0	ReferenceQueueDaemon
*7	14473	wait	2	1	FinalizerDaemon
*8	14474	timed-wait	0	1	FinalizerWatchdogDaemon
9	14475	native	0	0	Binder Thread #1
10	14476	native	0	0	Binder Thread #2
11	14479	running	22...	37	Thread-9238

Refresh　Fri Dec 07 15:25:47 CST 2012

Class	Method	File	Line	Native
com.shinado.shLauncher.ScrollView	run	ScrollView.java	221	false
java.lang.Thread	run	Thread.java	856	false

图 1-47　查看线程

查看内存分配

在进程栏中，选择要查看的进程，单击 Upgrade Heap 按钮 ，这个进程就会出现一个绿色小图标，如图 1-48 所示。

图 1-48　更新 heap 信息

同样，如图 1-49 所示，在右侧的窗口中选择 Allocation Tracker 栏，显示该进程下的内存使用情况。单击 Start Tracking 按钮，开始跟踪内存分配；单击 Get Allocation，显示结果。双击某一栏，下方会显示哪个地方产生这个 allocation。这个工具非常好用，特别是在进行游戏类的开发时，它会帮你分析你的瓶颈在哪里，帮你查看哪些地方可以加速，查看哪些变量可以重复使用。

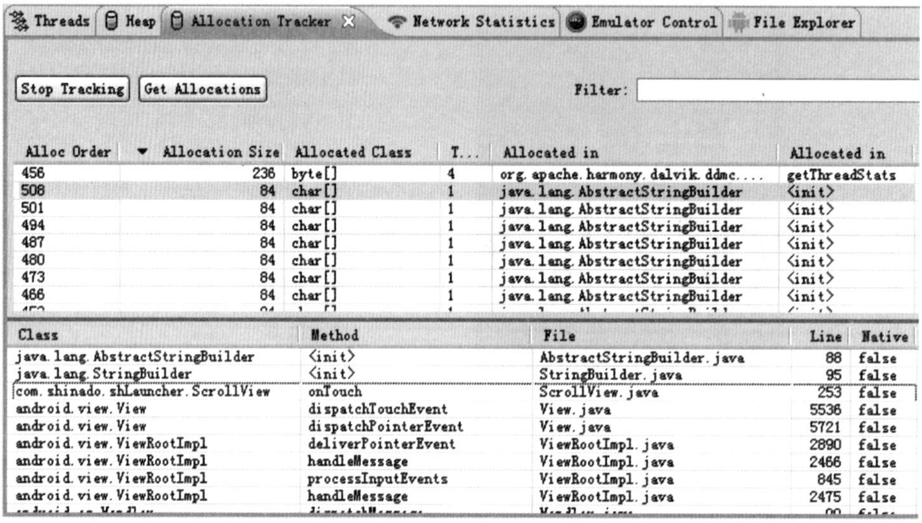

Threads | Heap | Allocation Tracker | Network Statistics | Emulator Control | File Explorer

Stop Tracking　Get Allocations　Filter:

Alloc Order	Allocation Size	Allocated Class	T...	Allocated in	Allocated in
456	236	byte[]	4	org.apache.harmony.dalvik.ddmc....	getThreadStats
508	84	char[]	1	java.lang.AbstractStringBuilder	<init>
501	84	char[]	1	java.lang.AbstractStringBuilder	<init>
494	84	char[]	1	java.lang.AbstractStringBuilder	<init>
487	84	char[]	1	java.lang.AbstractStringBuilder	<init>
480	84	char[]	1	java.lang.AbstractStringBuilder	<init>
473	84	char[]	1	java.lang.AbstractStringBuilder	<init>
466	84	char[]	1	java.lang.AbstractStringBuilder	<init>

Class	Method	File	Line	Native
java.lang.AbstractStringBuilder	<init>	AbstractStringBuilder.java	88	false
java.lang.StringBuilder	<init>	StringBuilder.java	95	false
com.shinado.shLauncher.ScrollView	onTouch	ScrollView.java	253	false
android.view.View	dispatchTouchEvent	View.java	5536	false
android.view.View	dispatchPointerEvent	View.java	5721	false
android.view.ViewRootImpl	deliverPointerEvent	ViewRootImpl.java	2890	false
android.view.ViewRootImpl	handleMessage	ViewRootImpl.java	2466	false
android.view.ViewRootImpl	processInputEvents	ViewRootImpl.java	845	false
android.view.ViewRootImpl	handleMessage	ViewRootImpl.java	2475	false

图 1-49　查看内存分配

第二章　Android基础

产品开发从这里开始

清单

Demo 代码：

```
\demo\Broadcast
\demo\ContentProvider
\demo\DrwableTest
\demo\HelloWorld
\demo\LayoutTest
\demo\WidgetTest
```

理论上来讲，学完这一章，你就可以做很多事情了。而且它不会花你太多的时间。抓住这几个要点：四大组件、布局、控件、Intent。

第一节　Android 四大组件

传说中的 Android 四大组件是 Activity，Service，Broadcast Receiver，Content Provider，这个问题在面试的时候就像 Activity 的生命周期，Android 的布局种类一样基础而重要。基本上，这几个问题可以判断你是不是知道 Android 开发。注意，是知道，甚至还没到了解的程度。

Activity

Activity 就是和用户交互的用户界面，你可以有点不准确地叫它“窗口”。在这个窗口中可以有很多个“控件”，也就是 Android 中的 View，比如说按钮、文本框、列表等。虽然在有些时候，View 可以直接而不通过 Activity 来跟用户进行交互(注意这一点，我们会在接下来的一章中专门提到)，但是通常来说，Activity 就是用户界面。也就是 MVC 框架中的 View 层。

我们每个用户窗口都要继承于 Activity，并在合适的时候重写它的生命周期方法。

常见异常

每个 Activity 都需要在 AndroidManifest. xml 文件中的 application 节点里声明，如：

```
<activity android:name = ". NewActivity"/>
```

否则会出现这样的异常：

```
AndroidRuntime  FATAL EXCEPTION: main
AndroidRuntime  java.lang.RuntimeException: Unable to start activity ComponentInfo{com.shin□
                ado.helloworld/com.shinado.helloworld.MainActivity}: android.content□
                ActivityNotFoundException: Unable to find explictit activity class{com□
                shinado.helloworld/com.shinado.helloworld.NewActivity}; have you declared□
                this active ty in your AndroidManifest.xml?
```

启动 Activity

还记得我们在前一章 AndroidManifest.xml 文件介绍中提到的 intent-filter 节点吗？当你打开这个应用时，系统就会打开你在 intent-filter 节点设置为 android.intent.action.MAIN 的 Activity。这个启动 Activity 的方法是由系统完成的，我们只需要把 Activity 的 intent-filter 节点设置为 android.intent.action.MAIN 就可以了。

Activity 生命周期

窗口是用来展示数据的，那么我们就必须得知道在什么时候把这些数据显示在窗口中，什么时候回收这些对象。因此就有了生命周期。Activity 的生命周期包括以下 7 个状态。

- onCreate：Activity 创建时调用。
- onStart：Activity 可见时调用。
- onResume：Activity 获得焦点时调用。
- onPause：Activity 可见但失去焦点时调用。
- onStop：Activity 不可见时调用。
- onDestroy：Activity 销毁时调用。
- onRestart：Activity 重新启动时调用。

我们的应用程序都是通过重写 onCreate 方法来设置初始值(初始化变量、设置监听器等)，而通过重写 onDestroy 方法做善后处理。这两个方法在 Activity 中异常重要。

它们的关系如图 2-1 所示。

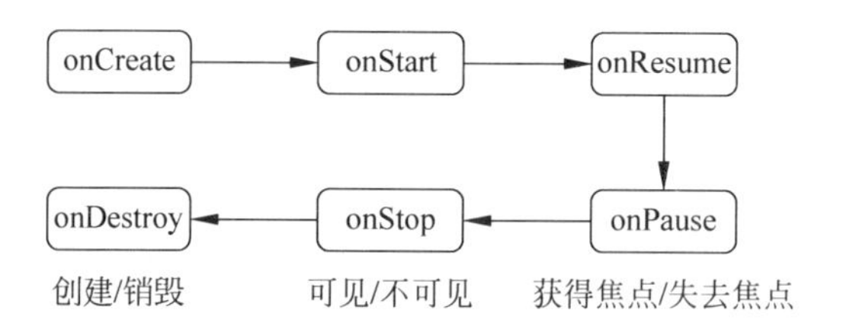

图 2-1　Activity 生命周期

我做了一个小试验，在 Activity 的每个生命周期中输出相应周期：

```
public class MainActivity extends Activity {

    private static String TAG = "HelloWorld";

    @Override
    public void onCreate(Bundle savedInstanceState) {
        super.onCreate(savedInstanceState);
```

```
            Log.i(TAG,"onCreate");
        }
        @Override
        public void onStart(){
            super.onStart();
            Log.i(TAG,"onStart");
        }
        @Override
        public void onResume(){
            super.onResume();
            Log.i(TAG,"onResume");
        }
        @Override
        public void onPause(){
            super.onPause();
            Log.i(TAG,"onPause");
        }
        @Override
        public void onStop(){
            super.onStop();
            Log.i(TAG,"onStop");
        }
        @Override
        public void onDestroy(){
            super.onDestroy();
            Log.i(TAG,"onDestroy");
        }
}
```

@Override 的重要性

大家可以看到每个重写的方法上面有个@Override 的标识，这个标识去掉之后也不会报错，但是当使用了@Override 的方法并没有重写该方法的话，就会报错。我在这里提醒大家，重写的时候尽量使用@Override 标识。因为有一次我重写 onCreate 方法时没有使用标识，把 onCreate 写成了 onCrate，找了半天硬是没有找到错误。如果有@Override 的标识的话，就可以避免类似的问题。

我做了四个动作，打开程序→返回主屏→回到程序→退出程序。程序的输出如图 2-2 所示。

Level	Time	Text	
I	01-06 20:03:30.879	onCreate	
I	01-06 20:03:30.879	onStart	
I	01-06 20:03:30.879	onResume	
I	01-06 20:03:38.785	onPause	← 返回主屏
I	01-06 20:03:45.761	onStop	
I	01-06 20:03:50.925	onStart	← 回到程序
I	01-06 20:03:50.925	onResume	
I	01-06 20:03:54.847	onPause	← 退出程序
I	01-06 20:03:55.320	onStop	
I	01-06 20:03:55.320	onDestroy	

图 2-2 不同操作下 Activity 的调用

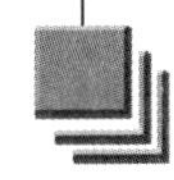

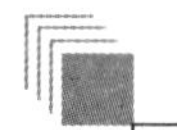

常见异常

重写 Activity 的生命周期时，如果没有调用父类方法，会抛出异常 SuperNotCall-Exception。如：

```
@Override
public void onCreate(Bundle savedInstanceState) {
    /**
     * 注释掉这行则抛出异常
     */
    //super.onCreate(savedInstanceState);
}
AndroidRuntime   FATAL EXCEPTION: main
AndroidRuntime   android.app.SuperNotCalledException: Activity{com.shinado.helloworld/com□
                 shinad o.helloworld.MainActivity}did not call through to super.onCreate()
```

Activity 非正常死亡

知识扩充

Android 系统会自动回收一些不需要的资源，来保证系统运行的流畅。所以以后不用再因为害怕自己手机内存不够而狂点"一键清除内存"之类的按钮。

不幸的是，Activity 可能会因为内存不够而被系统回收，这种事情一般发生在以下几种情况下：

- 按下 HOME 键，回到主屏时；
- 从一个 Activity 中跳到另一个 Activity 时；
- 按下电源按键(关闭屏幕)时；
- 长按 HOME 键，选择运行其他的程序时；
- 屏幕方向切换时。

这时，我们可以通过重写 onSaveInstanceState(Bundle savedInstanceState)方法来保存用户数据。当这个 Activity 被系统 kill 时，会先调用这个方法。只能说，Android 不是一个好的 killer，因为每次它都会给受害者留下遗言的时间(都是剧情需要)。还记得 onCreate 中的 Bundle 参数吗？当这个 Activity 被系统杀死然后被重新打开的时候，调用的 onCreate 方法中的 Bundle 参数就不会为空，而是一盘录音带，记录下了它临死前的所有数据(当然这些数据要程序员自己去添加)。

例如：

```
@Override
public void onCreate(Bundle savedInstanceState) {
    super.onCreate(savedInstanceState);
    /**
     * savedInstanceState 非空说明 Activity 被系统杀死，
     * onSaveInstanceState 方法被调用了
     */
    if(savedInstanceState != null){
        String key = savedInstanceState.getString("TEST_KEY");
    }
}
```

```
@Override
public void onSaveInstanceState(Bundle outState){
    super.onSaveInstanceState(outState);
    /**
      * 保存用户数据
      */
    outState.putString("TEST_KEY","sherlock");
}
```

小点子

下一次 HR 问你关于 Activity 生命周期的问题的时候，跟他讲这个。如果我是 HR 的话，我会对你印象深刻的。

Intent 和 Bundle

Intent 的中文意思是“意图”。你也可以这样理解，你可以通过 Intent 告诉各个组件你的意图，比如说你想要打开一个 Activity，比如你想要给正在收听某个特定“频道”广播的组件发送广播等等。Intent 的具体使用我们会在接下来的小节中分段介绍。

而 Bundle 相对好理解一点，它用于存放用户的数据。Intent 是桥梁，而 Bundle 是运输车，它能在组件之间运载东西给对方，如图 2-3 所示。

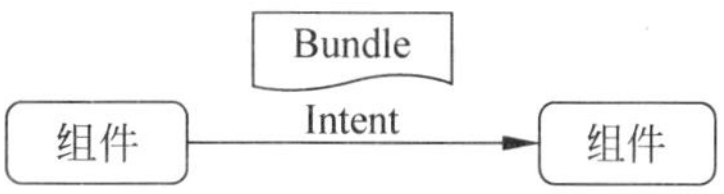

图 2-3　Intent 和 Bundle

Bundle 通过“键-值对”的方式存储数据，如：

```
Bundle bundle = new Bundle();
bundle.putString("TEST_KEY","sherlock");
```

通常来说，Bundle 由 Intent 携带，如：

```
intent.putExtra("EXTRA_BUNDLE",bundle);
```

如果你想要从一个 Activity 跳到另一个 Activity，Intent 就派上用场了：

```
Intent intent = new Intent();
intent.setClass(this,NewActivity.class);
//等同于
//Intent intent = new Intent(this,NewActivity.class);
startActivity(intent);
```

Service

Service 可以简单地理解为没有界面的 Activity，是后台运行的程序。音乐播放器是个很好的例子，当你退出播放界面回到主屏的时候，音乐会继续播放，自动播放下一曲。一般来说，Service 都是由 Activity 来启动（比如说，你打开音乐播放器（Activity），单击“播放”，开始音乐播放的 Service）。

声明 Service

首先，在工程的 src 文件夹上右击，选择 New→Package，填写包名新建一个 package，如图 2-4 所示。

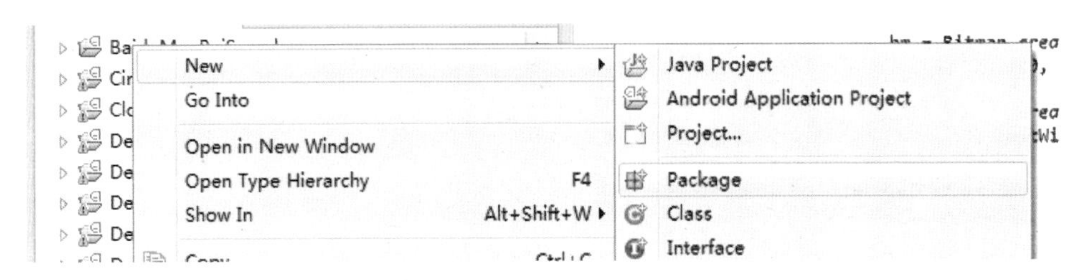

图 2-4　新建包

接下来，在这个新建的包上右击，选择 New→Class，如图 2-5 所示。

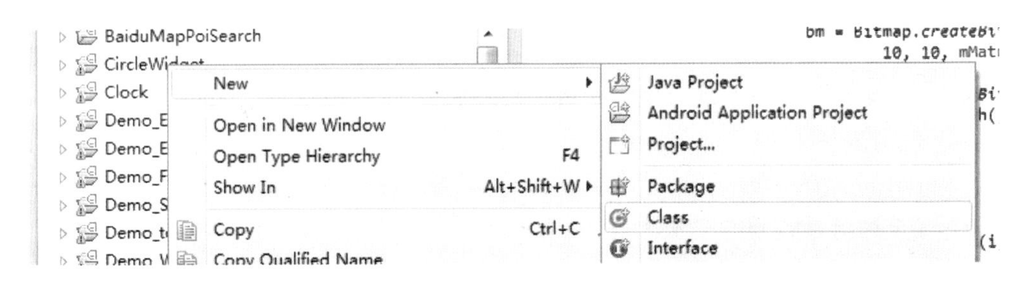

图 2-5　新建一个类

填写类名，新建一个 class，如图 2-6 所示。

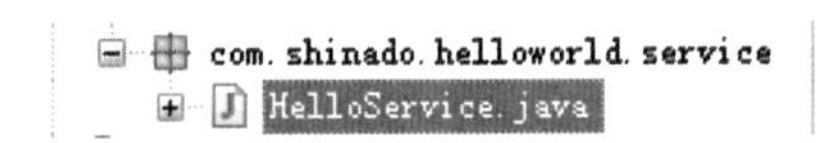

图 2-6　新建的类

初始的代码如下：

```
public class HelloService extends Service{
    @Override
    public IBinder onBind(Intent arg0) {
        //TODO Auto - generated method stub
        return null;
    }
}
```

跟 Activity 一样，Service 也必须在 AndroidManifest.xml 中声明，否则也不会报错，但是你会很迷茫为什么没有启动 Service：

```
<service android:name = ".service.HelloService"/>
```

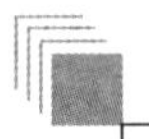

Service 生命周期

Service 有且仅有一个启动方法，那就是 Activity 的 startService 方法：

```
Intent intent = new Intent(NewActivity.this,HelloService.class);
NewActivity.this.startService(intent);
```

江湖中盛传的第二种启动 Service 的方法 bindService()并不能直接地理解为启动 Service。至于为什么要 bind Service，我们会在接下来的内容中讨论。

Service 的生命周期远比 Activity 简单，只有 onCreate、onStart 和 onDestroy 三种方法，如图 2-7 所示。

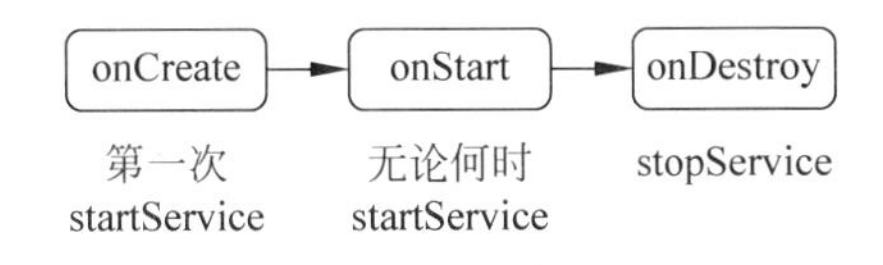

图 2-7 Service 生命周期

绑定 Service

我们回到音乐播放器的场景，用 activity 作为展示界面，activity 上面有各种按钮，而 service 控制音乐的播放、暂停、下一曲等。此时我在 activity 中单击了暂停按钮，那么我们要怎么让 service 去执行暂停的代码呢？换句话说，我们要怎么获取这个 service 对象，然后调用这个 service 对象的各种方法呢？

这时候我们需要 bindService(Intent service，ServiceConnection conn，int flags)这个方法：

- Service：要绑定的 service 意图。
- ServiceConnection：用于回调获取 service 对象。
- Flags：标识。

我们就可以通过 ServiceConnection 的回调方法来获取 service 对象，然后再调用 service 中的 play 方法。

```
private void bind(){
    Log.e(TAG,"在 Activity 中 bindService");
    Intent intent = new Intent(NewActivity.this,HelloService.class);
    NewActivity.this.startService(intent);
    NewActivity.this.bindService(intent,mConnection,0);
}

private void play() {
    service.play();
}
```

```
private HelloService service;
private ServiceConnection mConnection = new ServiceConnection() {
    //回调方法,当调用 bindService 时回调
    public void onServiceConnected(ComponentName className,
        IBinder localBinder) {
        Log.e(TAG,"在 Activity 中回调 onServiceConnected 方法");
        //获取 service 对象
        service = ((HelloBinder)localBinder).getService();
    }
    //回调方法,当调用 unbindService 时回调
    public void onServiceDisconnected(ComponentName arg0) {
        Log.e(TAG,"在 Activity 中回调 onServiceDisconnected 方法");
        service.onUnBindMethod();
        service = null;
    }
};
```

在 HelloService 中，在 onBind 方法中返回一个 HelloBinder 的对象，这个对象就是在 activity 中 onServiceConnected 方法的 localBinder。

```
private final IBinder binder = new HelloBinder();

@Override
public IBinder onBind(Intent arg0) {
    Log.e(TAG,"onBind");
    return binder;
}
public class HelloBinder extends Binder {
    public HelloService getService() {
        return HelloService.this;
    }
}
```

为了方便读者理解，我制作了一个小 demo(/demo/HelloWorld)。主要代码如下：

```
//按钮的监听器
private OnClickListener l = new OnClickListener(){
    public void onClick(View v) {
        int id = v.getId();
        Intent intent = new Intent(NewActivity.this,HelloService.class);
        switch(id){
        //单击 bind 按钮
        case R.id.activity_new_bind_bt:
            Log.e(TAG,"在 Activity 中单击 bind 按钮");
            NewActivity.this.bindService(intent,mConnection,0);
            break;
        //单击 unbind 按钮
        case R.id.activity_new_unbind_bt:
            Log.e(TAG,"在 Activity 中单击 unbind 按钮");
```

```
                NewActivity.this.unbindService(mConnection);
                break;
            //单击 start 按钮
            case R.id.activity_new_start_bt:
                Log.e(TAG,"在 Activity 中单击 start 按钮");
                NewActivity.this.startService(intent);
                break;
            //单击 stop 按钮
            case R.id.activity_new_stop_bt:
                Log.e(TAG,"在 Activity 中单击 stop 按钮");
                NewActivity.this.stopService(intent);
                break;
            //单击 play 按钮
            case R.id.activity_new_play_bt:
                Log.e(TAG,"在 Activity 中单击 play 按钮");
                service.play();
                break;
            }
        }
    };
```

当先后单击 start、bind 和 destroy 按钮时，输出如图 2-8 所示。

这时我们改变一下顺序，先单击 bind 再单击 start，然后是 unbind 和 destroy，得到的输入如图 2-9 所示。

```
Text
在Activity中点击start按钮
onCreate
onStart
在Activity中点击bind按钮
onBind
在Activity中回调onServiceConnected方法
Service调用onBindMethod方法
在Activity中点击stop按钮
在Activity中回调onServiceDisconnected方法
Service调用onUnBindMethod方法
onUnbind
onDestroy
```

图 2-8　Service 测试结果 1

```
Text
在Activity中点击bind按钮
在Activity中点击start按钮
onCreate
onBind
onStart
在Activity中回调onServiceConnected方法
Service调用onBindMethod方法
在Activity中点击unbind按钮
onUnbind
在Activity中点击stop按钮
onDestroy
```

图 2-9　Service 测试结果 2

我们可以看到，在 service 还未启动时单击 bind 按钮，不会调用 service 的任何方法，直到启动了 service 才会开始调用，并且顺序为 onCreate→onBind→onStart；而当我们先取消绑定 service 再终止 service 时，不会回调 onServiceDisconnectd 方法。

当然，类似的细节还有很多，笔者无法在这一一讲解，读者可以自己去探索。

Broadcast Receiver

顾名思义，Broadcast Receiver 就是广播接收者（收音机）。既然有收音机，那肯定是有广播的。在 Android 中，广播是无处不在的，如果你的程序也需要接受广播的话就必须使用 Broadcast Receiver。当然有广播就有特定的频道，我们可以在代码中设定自己喜欢的频道，接收有用的信息。

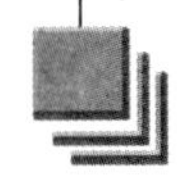

Android 系统中有很多系统设定的广播信息，比如电量广播。如果我们要做一个查看电量的程序(听起来有点多此一举)，那么就可以使用 Broadcast Receiver 监听电量广播。

再比如聊天系统的信息推送。我们使用一个 service 在后台接收消息，接收到消息后发送一个广播，告诉前台显示这一条消息。

发送和接收广播

先看一个简单的例子(/demo/Broadcast)，MainActivity 中启动服务 BroadcastService 并添加监听器，服务一启动就发送一个广播。

```
public class MainActivity extends Activity {
    private final static String TAG = "Broadcast";

    @Override
    public void onCreate(Bundle savedInstanceState) {
        super.onCreate(savedInstanceState);
        setContentView(R.layout.activity_main);

        //注册接收器,监听的 action 为 BroadcastService.ACTION
        IntentFilter filter = new IntentFilter();

        filter.addAction(BroadcastService.ACTION);
        registerReceiver(receiver,filter);

        //启动 service
        startService(new Intent(this,BroadcastService.class));
    }
    @Override
    public void onDestroy(){
        super.onDestroy();
        //注销这个接收器
        unregisterReceiver(receiver);
    }
    BroadcastReceiver receiver = new BroadcastReceiver(){
        @Override
        public void onReceive(Context context,Intent intent) {
            Bundle bundle = intent.getBundleExtra(
                BroadcastService.EXTRA_BUNDLE);
            String value = bundle.getString(BroadcastService.EXTRA_KEY);
            Log.e(TAG,value);
        }
    };
}
```

广播服务的代码：

```
public class BroadcastService extends Service{

    public static final String ACTION = "com.example.broadcast.test";
    public static final String EXTRA_BUNDLE = "bundle";
```

```
    public static final String EXTRA_KEY = "key";

    @Override
    public void onCreate(){
        super.onCreate();
        sendBroadcast();
    }
    private void sendBroadcast(){
        Intent intent = new Intent();
        intent.setAction(ACTION);
        Bundle bundle = new Bundle();
        bundle.putString(EXTRA_KEY,"sherlock");
        intent.putExtra(EXTRA_BUNDLE,bundle);
        sendBroadcast(intent);
    }
    @Override
    public IBinder onBind(Intent arg0) {
        //TODO Auto-generated method stub
        return null;
    }
}
```

注意代码中的黑体字，service 在发送广播的时候声明 action 为字符串 com.example.broadcast.test，可以把它当作一个特定的频道。而在 activity 中要接收的也正是这个频道（action 为字符串 com.example.broadcast.test）。

静态注册广播

新建一个类，继承 BroadcastReceiver：

```
public class TestBroadCastReceiver extends BroadcastReceiver{

    private final static String TAG = "TestBroadCastReceiver";

    @Override
    public void onReceive(Context context,Intent intent) {
        Bundle bundle = intent.getBundleExtra(
            BroadcastService.EXTRA_BUNDLE);
        String value = bundle.getString(BroadcastService.EXTRA_KEY);
        Log.e(TAG,value);
    }
}
```

然后在 Manifest.xml 的 application 节点下加入 receiver 节点：

```
<receiver android:name=".TestBroadCastReceiver">
    <intent-filter>
        <action android:name="com.example.broadcast.test"/>
    </intent-filter>
</receiver>
```

同样也能注册这个广播接收器。

Content Provider

Content Provider 多多少少是一个让人疑惑的组件，因为它不太常用，但是却归到了“四大组件”这个类别中。我们先来看看 Content Provider 的用途吧。首先它用于数据的存储和获取，也就是永久层。但是注意，它只是一套接口而已。也就是说具体怎么实现永久层，是使用 XML 存储还是 sqlite 存储或者网络存储，都由开发者自己实现。此外，Content Provider 还有一个特点，它支持不同程序的访问。也就是说，程序 B 可以访问到程序 A 的 Content Provider。

例如，我们可以在 Activity 中获取手机的所有联系人：

```
//获取联系人
private void getContacts() {
    //获取 ContentResolver 对象
    ContentResolver resolver = getContentResolver();
    //定义要查询的内容(姓名,号码和 id)
    String[] CONTACT_PROJECTION = new String[] {
        Phones.DISPLAY_NAME, Phones.NUMBER, Phones._ID };
    Cursor c = resolver.query(Phones.CONTENT_URI, CONTACT_PROJECTION, null, null, null);

    if (c != null) {
        while (c.moveToNext()) {
            //手机号码
            String number = c.getString(0);
            //联系人
            String name = c.getString(1);
            //id
            Long id = c.getLong(2);
            Log.e(TAG, id + ":" + name + " " + number);
        }
        c.close();
    }
}
```

第二节　Android UI 设计

在这一节，我们可以先抛下枯燥无聊的代码（如果你喜欢代码的话那另当别论），来看一些赏心悦目的东西。

设计规范

2012 年初，Android 终于退出官方界面设计指导网站，有兴趣的可以上网看一下（http://developer.android.com/design/index.html）。笔者结合自己的经验和官方指导网

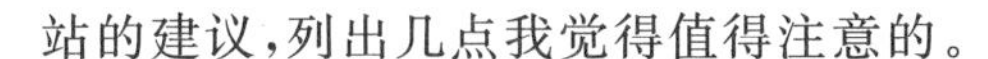

站的建议，列出几点我觉得值得注意的。

要有反馈，特别是按钮

Android 的官方说明是，按钮至少要有图 2-10 所示的四种状态（默认、不可用、获得焦点、按下）。我们经常会忘记获得焦点和不可用这两种状态，但是随着触摸笔技术以及智能电视的普及，应该要开始注意获得焦点这个状态的处理。

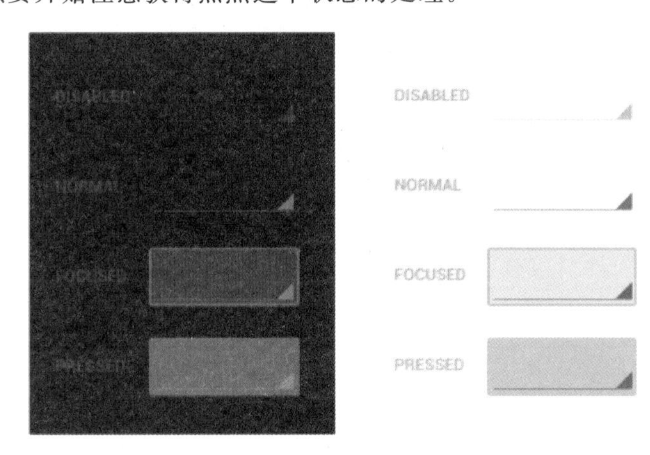

图 2-10　按钮四种状态

按钮能按的区域要大于按钮本身

这一点你应该很有经验，因为我们说不准自己的手指到底按在什么地方了。

让用户惊讶，不要让用户震惊

提供最简单最直接的操作方式，行有余力之时再给用户提供便捷操作。比如在应用 todo（早期版本）中，有向右滑动删除事件的功能（图 2-11），但同时也有通过单击按钮删除事件的功能。如果没有常规的按钮删除功能，用户估计会抓狂的。

图 2-11　滑动操作

用图片说话

如果有非常规操作，用引导图来告诉用户怎么做。比如在应用 IxLaunch 中，当用户打开程序后，会有图片引导，直到用户学会怎么操作，如图 2-12 所示。

突出重点

在课程表应用中，整个界面的重点是查看，因为使用课程表最重要的功能是查看课程；而每节课的重点是时间，谁都不想迟到。让重点突出，一目了然，如图 2-13 所示。

图 2-12　图片引导

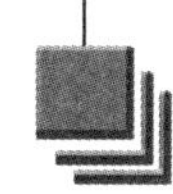

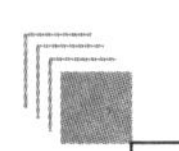

图 2-13　突出重点

隐藏非重要信息

作为一个合格的菜单，它的必备素质之一就是知道怎么隐藏。在控件应用中，当菜单不使用时隐藏在左下角，需要时弹出，如图 2-14 所示。

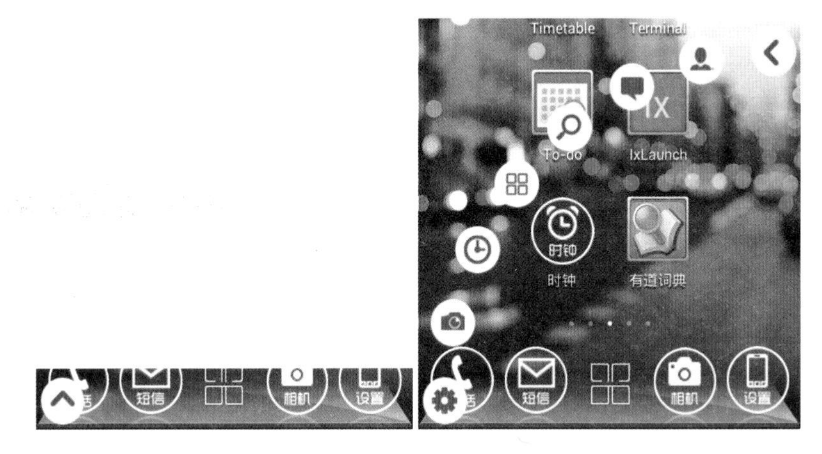

图 2-14　需要时弹出

多用 sytle

减少代码冗余。

不要使用 px，使用 dp

这一点会在接下来说明。

不要使用 AbsoluteLayout

虽然绝对布局是客观存在的，但是不要用它，就像虽然有 goto 但是没有人会用它一样。

屏幕适应

作为一个 Android 开发者，屏幕的事情总是让人头疼。不过好在 Android 有自己的一套兼容性方案。但是读者在开发 Android 的过程中需要注意这些问题。

drawable 文件夹

前面一章中介绍过，drawable 文件夹有四个：drawable-xhdpi、drawable-hdpi、drawable-mdpi 和 drawable-ldpi。我们新建一个工程，将布局改为一张居中的图片，图片用的就是生成工程时自带的图标。

```
<RelativeLayout
    xmlns:android = "http://schemas.android.com/apk/res/android"
    xmlns:tools = "http://schemas.android.com/tools"
    android:layout_width = "fill_parent"
    android:layout_height = "fill_parent" >
    <ImageView
        android:layout_width = "wrap_content"
        android:layout_height = "wrap_content"
        android:layout_centerHorizontal = "true"
        android:layout_centerVertical = "true"
        android:src = "@drawable/ic_launcher"/>
</RelativeLayout>
```

我用了三个不同分辨率的模拟器来运行这个程序，得到的结果如图 2-15 所示（左边 800×480，右上 320×240，右下 480×320）。虽然分辨率不同，但是显示效果是一样的。这是因为在每个 drawable 文件夹中有相对应尺寸的图片，分别是 72×72，48×48 和 32×32。

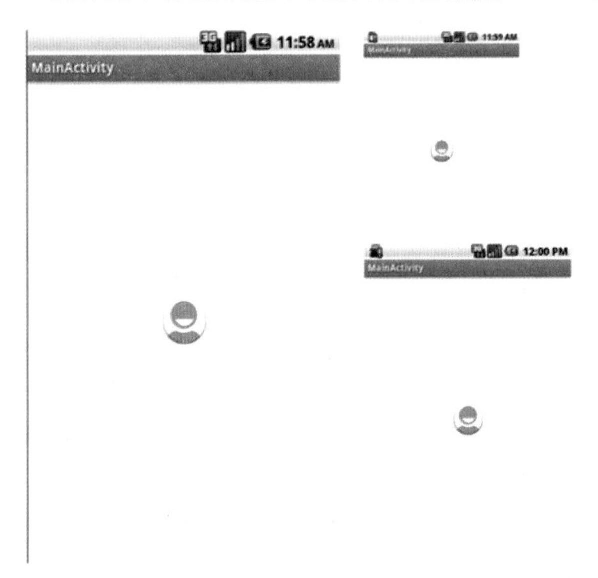

图 2-15　不同分辨率下的 ImageView

我们再做一个试验，保留 drawable-ldpi 文件夹，删除其他 drawable 文件夹，然后再运行一次（注意要修改代码文件，随便加个空格都可以，否则 ADT 不会重新编译）。此时高分辨率的模拟器上的图片呈现模糊状态，如图 2-16 所示因为图片被拉伸至它应该有的大小。

图 2-16　高分辨率下图片被拉伸

最后，将 drawable-ldpi 重命名为 drawable，再次运行。这时所有的图标都会缩小，如图 2-17 所示。

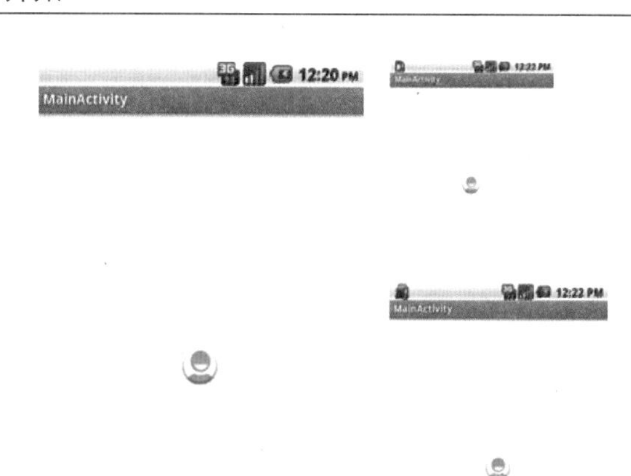

图 2-17　所有分辨率都使用同一个图片

长度单位 dp 和 px

先了解一下 dp 和 px 这两个概念。dp 等同于 dip(device-independent pixel)，也就是跟设备分辨率无关的像素。Android 把手机设备分为 480dip×320dip；px 也就是 pixel 像素，跟屏幕分辨率有关。

为了更具体地演示 dp 和 px 的区别，我们再做一下修改，把布局文件的图片框代码改为一个 80 像素×80 像素的黑色矩形：

```
<ImageView
    android:layout_width="80px"
    android:layout_height="80px"
    android:layout_centerHorizontal="true"
    android:layout_centerVertical="true"
    android:src="@android:color/black"/>
```

在不同的分辨率下，这个黑色矩形保持相同的物理大小，如图 2-18 所示。

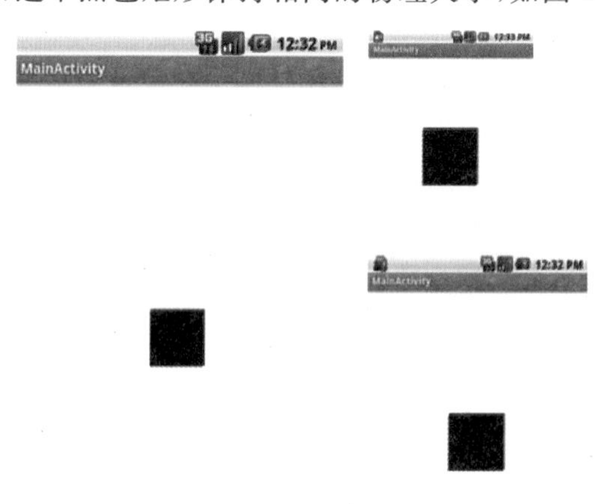

图 2-18　不同分辨率下矩形物理大小相同

再把长和宽都改成 80dp，就回到了“看起来一样大”的效果，如图 2-19 所示。

```
android:layout_width = "80dp"
android:layout_height = "80dp"
```

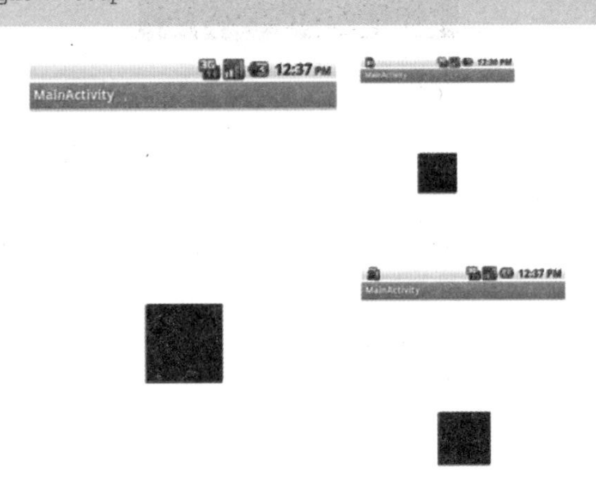

图 2-19　不同分辨率下矩形相对大小相同

使用 9png

在 Android 开发中，经常使用 9png 格式的图片作为控件的背景，它的好处在于可以控制缩放的区域，达到缩放自如的效果，如图 2-20 所示。

图 2-20　使用 9png 的效果

9png 的原理就是指定缩放图片中的部分，在这个输入框背景中，指定缩放的部分为圆角内的方形部分，从而不会影响到圆角的图形。

最新的 Android SDK 中自带了 9png 图片的制作工具，在\sdk\tools 中可以找到该工具 draw9patch. bat，打开后的界面如图 2-21 所示。

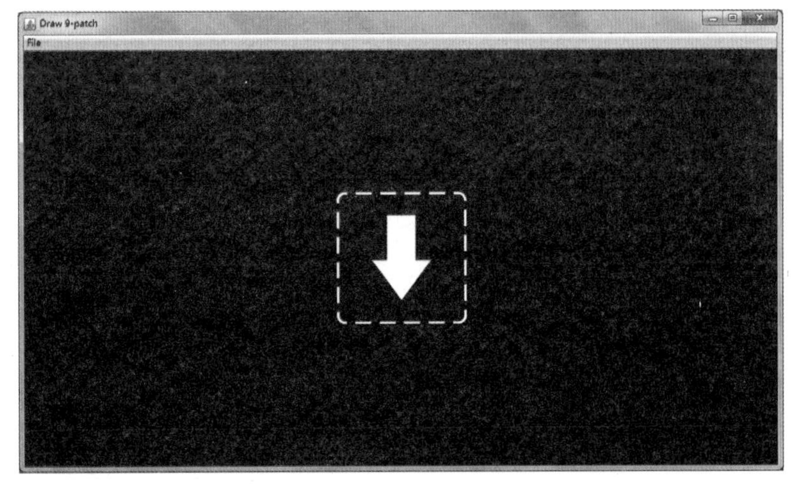

图 2-21　9png 工具

将需要制作的 png 图片拖动到上面。我们以图 2-20 为例。

首先在界面下方勾选 Show content 和 Show patches 以便 9png 的制作，如图 2-22 所示。

图 2-22　勾选 Show content 和 Show patches

在图片四周各点一个点，行成一个十字，如图 2-23 所示。

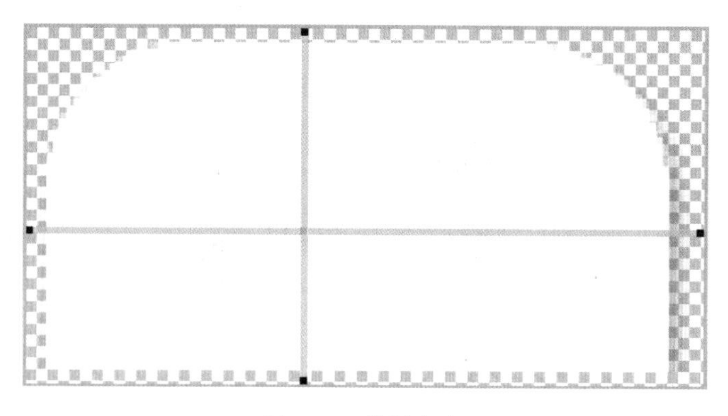

图 2-23　绘制十字

此时，在右图的拉伸效果示意中已经出现 9png 的雏形，但这只是第一步。图中阴影区域表示可以显示的区域，如图 2-24 所示。

此时在手机中的显示效果为如图 2-25 所示。

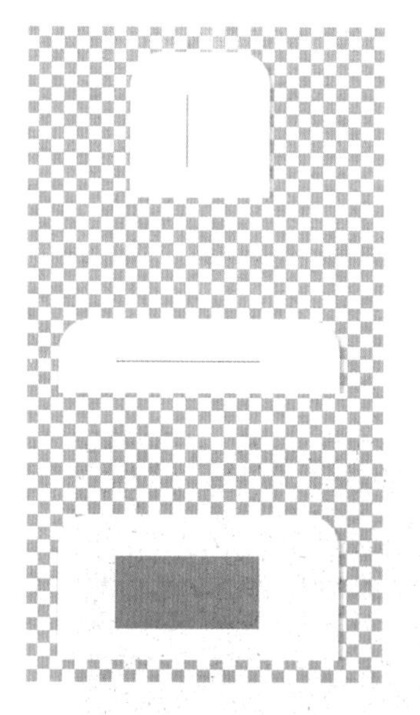

图 2-24　显示区域

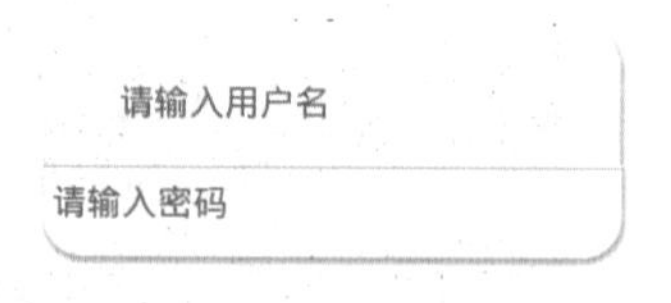

图 2-25　实际效果

纠正显示区域，补上两边的线，如图 2-26 所示。

这个 9png 图片就做好了。

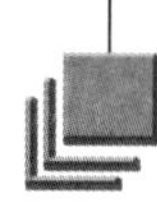

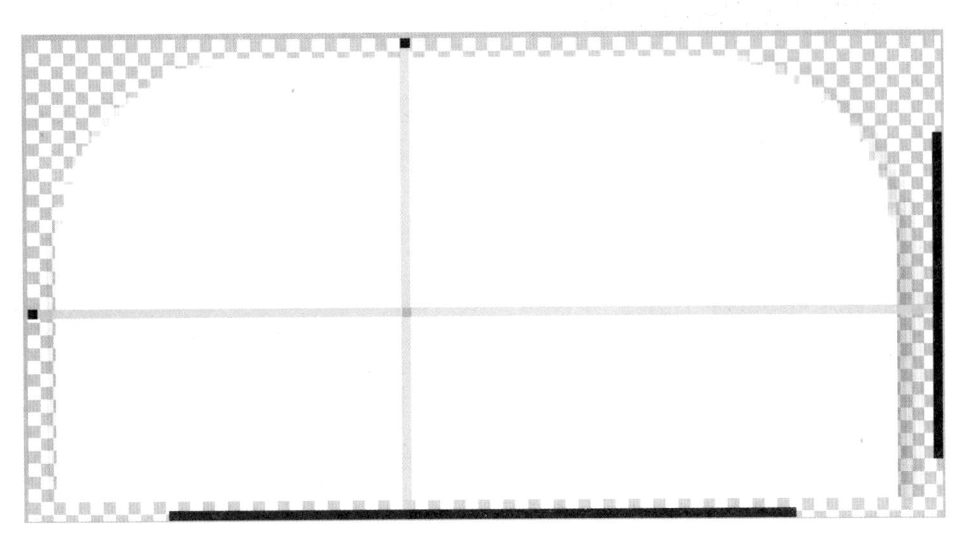

图 2-26　纠正显示区域

你的风格？

现在比较流行并且也是较为对立的两种设计风格是以苹果为代表的 Skeuomorph（拟物）风格和以微软为代表的 Metro 风格（图 2-27，成书后不久，苹果公司也加入到扁平化的大潮中）。

图 2-27　Metro 风格

我个人偏爱类似 Metro 的简洁风格，因为这种风格的应用太好做了，在没有专业美工的情况下也能做出很好看的应用（偷笑）。你只用把文字、图片放到正确的地方，调一个正确的大小，看起来错落有致，简洁有力就可以了。图 2-28 就是简洁风络的一个例子。

当然拟物风格也无妨，课程表中的概念时钟控件（我打赌你看不出它是一个时钟）就是一个拟物的例子，如图 2-29 所示随着指针的调动，会发出“吱吱”的机械声音，非常好玩。

图 2-28　清新风格界面

图 2-29　拟物风格

第三节　Android 五大布局

我们首先新建一个布局文件，在 layout 文件夹上右击，选择 New→Android XML File，如图 2-30 所示。

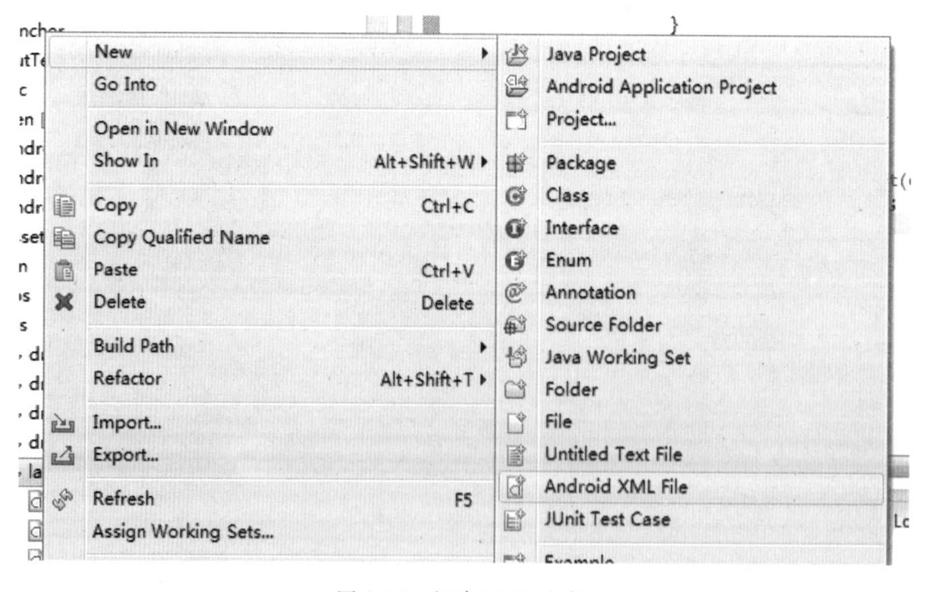

图 2-30　新建 XML 文件

在接下来的界面中填写文件名（注意，要以. xml 结尾），选择根元素，如图 2-31 所示。

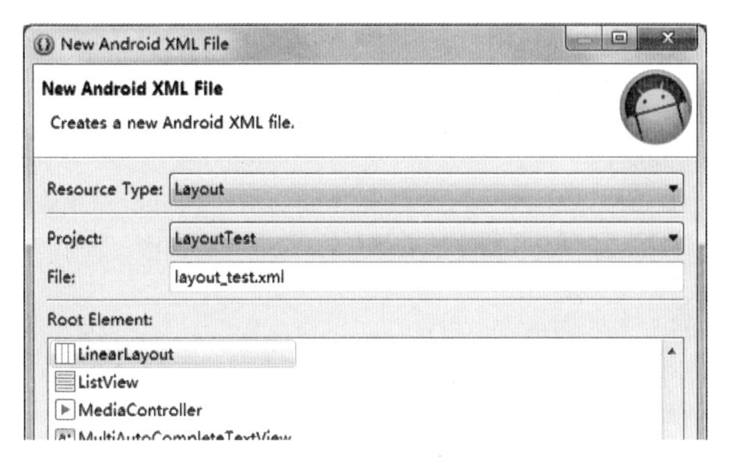

图 2-31　选择根元素

我们选择 LinearLayout，并且修改代码：

```
<?xml version = "1.0" encoding = "utf - 8"?>
< LinearLayout
    xmlns:android = "http://schemas.android.com/apk/res/android"
    android:layout_width = "fill_parent"
    android:layout_height = "fill_parent">

    < Button
        android:layout_width = "wrap_content"
        android:layout_height = "wrap_content"
        android:text = "@string/hello_world"/>

</LinearLayout >
```

这个布局文件包括一父一子两个节点，父节点为 LinearLayout 线性布局，它是一个容器，用于盛放各种控件，当然也包括它自己；子节点为一个文本。在 Android 中，可以盛放子容器的布局有五种，接下来我们会一一介绍。

Android 4.0 以后就已经是六大布局了

Android 4.0 以后增加了第六个布局 GridLayout，会在第 9 章详细介绍。

万金油 AbsoluteLayout 绝对布局

绝对布局，也就是根据坐标确定控件的布局方法，现在的 ADT 已经把它列为“弃用”了。虽然如此，但是在奇葩的情况下也是会有使用的。

我们新建一个布局文件 layout_absolute.xml，选择根元素为 AbsoluteLayout。

创建完成之后，layout_absolute.xml 文件会包含根元素 AbsoluteLayout：

```
<?xml version = "1.0" encoding = "utf - 8"?>
< AbsoluteLayout
    xmlns:android = "http://schemas.android.com/apk/res/android"
```

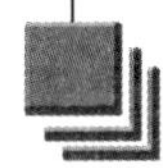

```
    android:layout_width = "fill_parent"
    android:layout_height = "fill_parent" >

</AbsoluteLayout >
```

这里两个属性 android:layout_width 和 android:layout_height 分别表示这个元素的宽度和长度，这两个属性是几乎所有控件都必须有的，没有就会报错。它的可填参数如下。

- fill_parent：扩充至父元素，与父元素大小相同。
- wrap_content：与内容的大小相同。

最新版的 ADT 已经可以很好地支持用图形界面来创建布局文件了。通过切换下面的卡片来选择图形界面和代码界面，如图 2-32 所示。

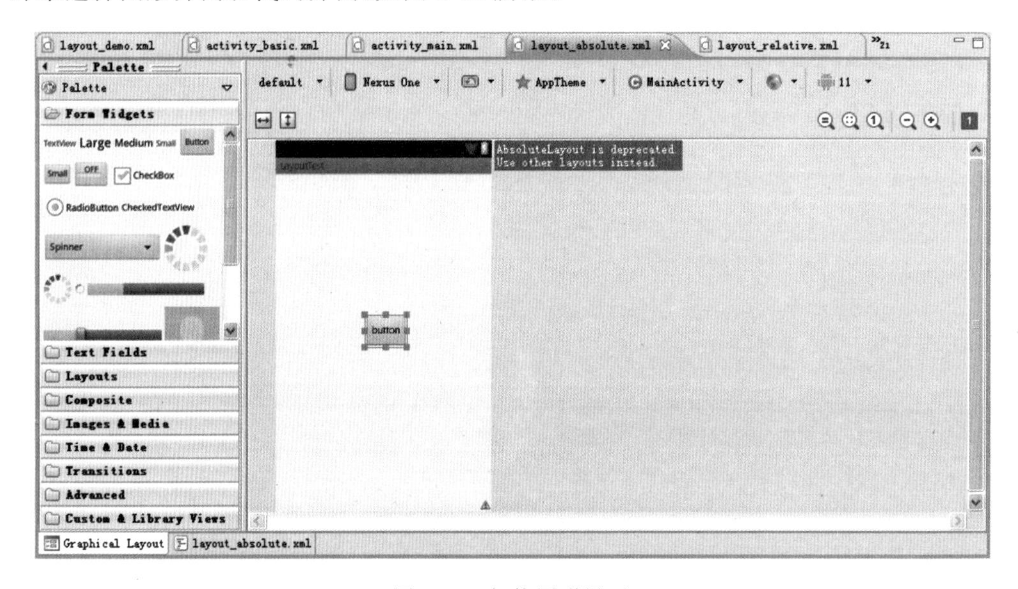

图 2-32　切换图形界面

把一个 Butto 拖到右边的布局中，会自动产生代码：

```
< Button
        android:id = "@ + id/button1"
        android:layout_width = "wrap_content"
        android:layout_height = "wrap_content"
        android:layout_x = "134dp"
        android:layout_y = "202dp"
        android:text = "@string/button"/>
```

作为初学者来说，建议读者朋友们不要使用图形界面来创建布局文件，而要自己写一写代码。因为如果一开始就使用图形界面就会产生一种依赖，有时候出了错也不知道是哪里出错。建议读者在熟悉了布局文件以及各个控件之后再来使用图形化的界面。

言归正传，绝对布局的特征就是子控件有两个属性：android:laytou_x 和 android:layout_y。顾名思义，它们分别表示控件的 x 坐标和 y 坐标。

最后说明一句，慎用绝对布局。

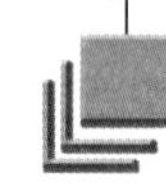

聪明的 RelativeLayout 相对布局

相对布局是 Android 布局中较为灵活的一种，我们可以通过设置控件相对于参照物的相对位置(比如说按钮 1 在编辑框 1 的下方)来设置布局。

我们来看一个相对布局的例子。这个布局有 5 个 Button，分别为 center、left、right、top 和 bottom。顾名思义，它们的位置分别在中心、左边、右边、上方和下方，如图 2-33 所示。

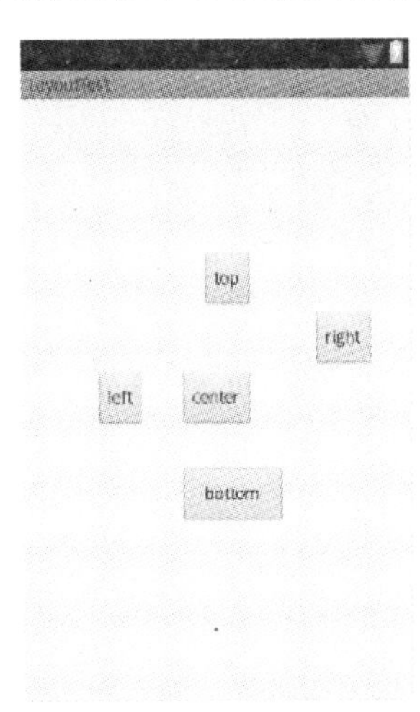

图 2-33 相对布局实例

布局文件 layout_ralative.xml 代码为：

```
<?xml version="1.0" encoding="utf-8"?>
<RelativeLayout xmlns:android="http://schemas.android.com/apk/res/android"
    android:layout_width="fill_parent"
    android:layout_height="fill_parent" >

    <Button
        android:id="@+id/center"
        android:layout_width="wrap_content"
        android:layout_height="wrap_content"
        android:layout_centerHorizontal="true"
        android:layout_centerVertical="true"
        android:text="@string/center"/>

    <Button
        android:id="@+id/bottom"
        android:layout_width="wrap_content"
        android:layout_height="wrap_content"
        android:layout_alignLeft="@+id/center"
        android:layout_below="@+id/center"
        android:layout_marginTop="30dp"
        android:text="@string/bottom"/>

    <Button
        android:id="@+id/right"
```

```
        android:layout_width="wrap_content"
        android:layout_height="wrap_content"
        android:layout_above="@+id/center"
        android:layout_marginLeft="22dp"
        android:layout_toRightOf="@+id/bottom"
        android:text="@string/right"/>

    <Button
        android:id="@+id/left"
        android:layout_width="wrap_content"
        android:layout_height="wrap_content"
        android:layout_above="@+id/bottom"
        android:layout_marginRight="28dp"
        android:layout_toLeftOf="@+id/center"
        android:text="@string/left"/>

    <Button
        android:id="@+id/top"
        android:layout_width="wrap_content"
        android:layout_height="wrap_content"
        android:layout_above="@+id/right"
        android:layout_alignRight="@+id/center"
        android:text="@string/top"/>

</RelativeLayout>
```

我们先来看第一个按钮 center，它有两个特殊属性：

- android:layout_centerHorizontal：设置为 true 时，控件水平居中。
- android:layout_centerVertical：设置为 true 时，控件垂直居中。

因此 center 按钮呈水平居中。除此之外，在绝对布局中还有另外四个属性用于设置没有参照物的控件的位置：

- android:layout_alignParentTop：当设置为 true 时，控件靠父容器的上方。
- android:layout_alignParentBottom：当设置为 true 时，控件靠父容器的下方。
- android:layout_alignParentLeft：当设置为 true 时，控件靠父容器的左边。
- android:layout_alignParentRight：当设置为 true 时，控件靠父容器的右边。

我们以按钮 center 为参照物，就有按钮 bottom；它有以下三个属性：

- android:layout_alignLeft：该控件与指定控件的左边对齐，格式为"@+id/[控件 id]"。
- android:layout_below：该控件在指定控件的下方，格式为"@+id/[控件 id]"。
- android:layout_marginTop：像素，表示该控件与指定的上方控件的距离。

同样，以 center 和 bottom 为参照物，有按钮 right，同样，它也有以下三个属性：

- android:layout_above：该控件在指定控件的上方，格式为"@+id/[控件 id]"。
- android:layout_marginLeft：像素，表示该控件与指定的左边控件的距离。
- android:layout_toRightOf：该控件在指定控件的右边，格式为"@+id/[控件 id]"。

乍一看，似乎 android:layout_above 和 android:layout_toRightOf 不是表示同一个格式

的属性(在潜意识中,与 android:layout_above 相对应,表示在指定控件的右边应该为“android:layout_right”)。但实际上相对布局中是没有 android:layout_right 这个属性的,因为英语中的语法是 A is to right of B,而不是 A is right B。

同样,top 跟 left 按钮的布局也是以类似的方法设置的,这里就不多加赘述了。

总结一下,设置控件与参照物的相对位置的常用方法有 12 种,分别如下:

- android:layout_top:该控件在指定控件的上方,格式为"@+id/[控件 id]"。
- android:layout_below:该控件在指定控件的下方,格式为"@+id/[控件 id]"。
- android:layout_toLeftOf:该控件在指定控件的左边,格式为"@+id/[控件 id]"。
- android:layout_toRightOf:该控件在指定控件的右边,格式为"@+id/[控件 id]"。
- android:layout_alignTop:该控件与指定控件的上方对齐,格式为"@+id/[控件 id]"。
- android:layout_alignBelow:该控件与指定控件的下方对齐,格式为"@+id/[控件 id]"。
- android:layout_alignLeft:该控件与指定控件的左边对齐,格式为"@+id/[控件 id]"。
- android:layout_alignRight:该控件与指定控件的右边对齐,格式为"@+id/[控件 id]"。

易扩展的 LinearLayout 线性布局

线性布局是 Android 布局中使用最多的布局,使用起来也很方便,子控件中没有特殊的属性,只需要嵌套在 LinearLayout 元素中即可。

我们先来看一个线性布局的例子。它包含了有线性布局嵌套线性布局的结构,如图 2-34 所示。

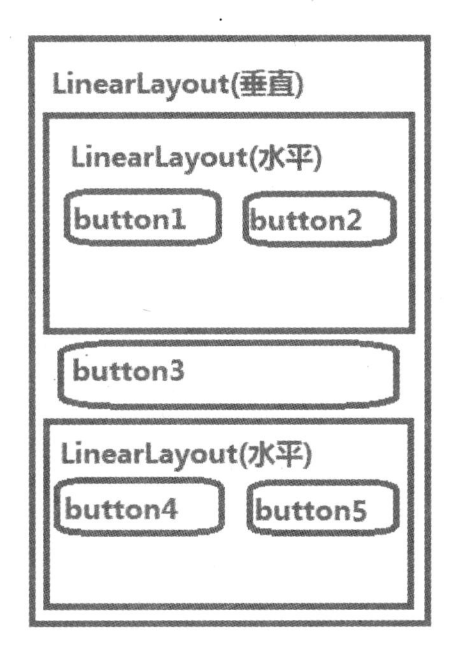

图 2-34　布局示意图

这个布局文件的第一层为垂直 LinearLayout,第二层有两个水平 LinearLayout 和一个 Button(button3),第三层为 4 个 Button,分别在两个 LinearLayout 之中。理解了这个结构之后我们再来看布局文件:

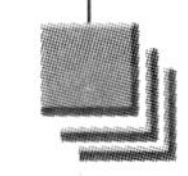

```xml
<?xml version = "1.0" encoding = "utf - 8"?>
<LinearLayout
    xmlns:android = "http://schemas.android.com/apk/res/android"
    android:layout_width = "fill_parent"
    android:layout_height = "fill_parent"
    android:orientation = "vertical" >

    <LinearLayout
        android:layout_width = "fill_parent"
        android:layout_height = "wrap_content" >

        <Button
            android:id = "@+ id/button1"
            android:layout_width = "wrap_content"
            android:layout_height = "wrap_content"
            android:text = "@string/left"/>

        <Button
            android:id = "@+ id/button2"
            android:layout_width = "wrap_content"
            android:layout_height = "wrap_content"
            android:text = "@string/right"/>

    </LinearLayout>

    <Button
        android:id = "@+ id/button3"
        android:layout_width = "wrap_content"
        android:layout_height = "wrap_content"
        android:text = "@string/center"
        />

    <LinearLayout
        android:layout_width = "fill_parent"
        android:layout_height = "wrap_content" >

        <Button
            android:id = "@+ id/button4"
            android:layout_width = "wrap_content"
            android:layout_height = "wrap_content"
            android:text = "@string/bottom"/>

        <Button
            android:id = "@+ id/button5"
            android:layout_width = "wrap_content"
            android:layout_height = "wrap_content"
            android:text = "@string/bottom"/>

    </LinearLayout>
</LinearLayout>
```

运行结果如图 2-35 所示。

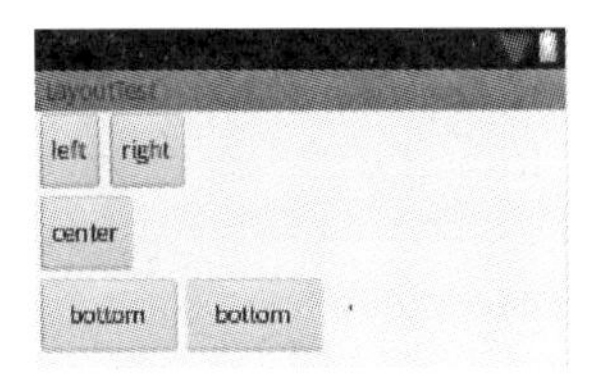

图 2-35　线性布局实例

切换 FrameLayout 卡片布局

卡片布局可以说是 Android 中最傻瓜式的布局了。卡片布局中每个控件的左上方都与父容器(也就是卡片布局本书)左上方重合,在父容器中层叠起来。

说起来略显抽象。我们来看一个例子就会明白卡片布局的作用:

```
<?xml version = "1.0" encoding = "utf - 8"?>
< FrameLayout xmlns:android = "http://schemas.android.com/apk/res/android"
    android:layout_width = "fill_parent"
    android:layout_height = "fill_parent" >

    < TextView
        android:id = "@ + id/textView1"
        android:layout_width = "wrap_content"
        android:layout_height = "wrap_content"
        android:text = "@string/hello_world"
        android:textAppearance = "?android:attr/textAppearanceLarge"/>

    < Button
        android:id = "@ + id/button1"
        android:layout_width = "wrap_content"
        android:layout_height = "wrap_content"
        android:text = "@string/button"/>

    < ImageView
        android:id = "@ + id/imageView1"
        android:layout_width = "wrap_content"
        android:layout_height = "wrap_content"
        android:src = "@drawable/ic_launcher"
        android:contentDescription = "@string/hello_world"/>

</FrameLayout >
```

运行结果如图 2-36 所示。

图 2-36　卡片布局实例

表单 TableLayout 表格布局

表格布局最经典的应用就是用户注册。表格布局必须包含子控件＜TableRow＞,它表示表格中的一行,一行中又可以包含多个控件(水平),各个控件在水平上相互对齐。可以说,表格布局就是一个垂直 LinearLayout 嵌套多个水平 LinearLayout。

我们来看一个例子：

```
<?xml version="1.0" encoding="utf-8"?>
<TableLayout xmlns:android="http://schemas.android.com/apk/res/android"
    android:layout_width="fill_parent"
    android:layout_height="fill_parent" >

    <TableRow
        android:layout_height="wrap_content"
         android:gravity="center" >

        <TextView
            android:layout_width="wrap_content"
            android:layout_height="wrap_content"
            android:text="@string/user_name"/>

        <EditText
            android:layout_width="150dp"
            android:layout_height="wrap_content"
            android:inputType="text"/>

    </TableRow>

    <TableRow
        android:layout_height="wrap_content"
         android:gravity="center" >

        <TextView
            android:layout_width="wrap_content"
            android:layout_height="wrap_content"
            android:text="@string/pwd"/>

        <EditText
            android:layout_width="150dp"
            android:layout_height="wrap_content"
            android:inputType="textPassword"/>

    </TableRow>

    <TableRow
        android:layout_height="wrap_content"
         android:gravity="center" >

        <TextView
            android:layout_width="wrap_content"
            android:layout_height="wrap_content"
            android:text="@string/pwd_rep"/>

        <EditText
            android:layout_width="150dp"
```

```
            android:layout_height = "wrap_content"
            android:inputType = "textPassword"/>
    </TableRow>

</TableLayout>
```

如果读者细心的话就会发现 TableRow 居然没有 android:layout_width 这个属性。我们在之前说过 android:layout_width 这个属性是几乎所有控件都必须有的，这个 TableRow 就是个特例了。因为 TableRow 的 android:layout_width 默认为“fill_parent”，而且不能修改，所以我们可以不设置，即便设置了其他参数也没有效果。

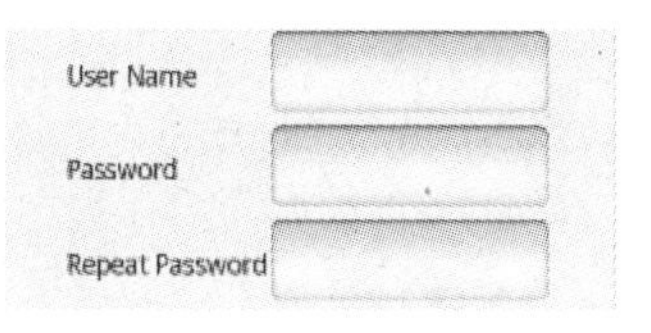

图 2-37 表格布局实例

如图 2-37 所示，我们可以看到，每一列的按钮长度都是一样的，也就是说每一列的控件长度取决于这一列长度最长的控件。

惊！用 Java 代码也能配置布局？

除了常规地使用布局文件配置布局之外，也可以在 Java 代码中配置布局，虽然这样不符合 Android 的代码规范，但是在某些特定情况下，比如一个界面的按钮依上下文而定，是会使用这种方式来创建 activity 的视图的。

一个简单的代码示例如下：

```
@Override
public void onCreate(Bundle savedInstanceState){
    super.onCreate(savedInstanceState);

    LinearLayout root = new LinearLayout(this);
    //android:orientation = "vertical"
    root.setOrientation(LinearLayout.VERTICAL);

    /*
     * android:layout_width = "fill_parent"
       android:layout_height = "fill_parent"
     */
    LinearLayout.LayoutParams params = new LinearLayout.LayoutParams(
                LinearLayout.LayoutParams.FILL_PARENT,
                LinearLayout.LayoutParams.FILL_PARENT);
    root.setLayoutParams(params);

    //两个 button
    Button bt_1 = new Button(this);
    bt_1.setText("button 1");
    root.addView(bt_1);
    Button bt_2 = new Button(this);
    bt_2.setText("button 2");
    root.addView(bt_2);
```

```
        //设置视图
        setContentView(root);
    }
```

Android 中的五个布局就在这里讲完了。但是关于布局的学习还不止如此，笔者介绍的这些只能使用在没有特殊情况下的布局中。当有特殊情况时，读者就要自己去探究学习。任何学科的学习大抵如此，所谓“师傅领进门，修行在个人”。

第四节　Android 基本控件

这一节我们会介绍 Android 的基本控件，但是更重要的是我们要知道怎么自定义，怎么美化。所以，忘了它们原来的样子吧。本节以线性布局为根布局，讲解 Android 基本控件。

公共控件属性

在控件属性中，有几个是非常基础的，每个控件（包括布局）都会有这个属性。在这里详细讲解 Android 控件中常见的属性。

layout_gravity 和 gravity

我们先看一个简单的例子：

```
<?xml version = "1.0" encoding = "utf - 8"?>
<LinearLayout xmlns:android = "http://schemas.android.com/apk/res/android"
    android:layout_width = "fill_parent"
    android:layout_height = "fill_parent"
    android:gravity = "center_vertical"
    android:orientation = "vertical" >

    <Button
        android:layout_gravity = "right"
        android:gravity = "left"
        android:layout_width = "200dp"
        android:layout_height = "50dp"
        android:text = "@string/button"
        />

</LinearLayout>
```

我们会发现这个按钮垂直居中于整个屏幕，水平靠左；按钮内的文字位置为左上角。先来分析 LinearLayout 的设置：

```
android:gravity = "center_vertical"
```

这表明它的子节点（在这里也就是按钮）相对于自己垂直居中。

再来看 Button 的设置：

```
android:layout_gravity = "right"
```

这表明它相对于父节点(也就是 LinearLayout),位置为右边。

```
android:gravity = "left"
```

这表明它的子元素(也就是文字)相对于自己靠左排列,如图 2-38 所示。

图 2-38　gravity 属性设置结果

padding 和 margin

先来看个例子,显示效果如图 2-39 所示。

```
<?xml version = "1.0" encoding = "utf - 8"?>
< LinearLayout xmlns:android = "http://schemas.android.com/apk/res/android"
    android:layout_width = "fill_parent"
    android:layout_height = "fill_parent"
    android:orientation = "vertical" >

    < Button
        android:paddingTop = "24dp"
        android:paddingLeft = "72dp"
        android:layout_marginLeft = "24dp"
        android:layout_marginTop = "72dp"
        android:layout_width = "wrap_content"
        android:layout_height = "wrap_content"
        android:text = "@string/button"
        android:textSize = "30dp"/>

</LinearLayout >
```

padding 和 margin 很容易被混淆。只要记住 padding 是对子元素,而 margin 是相对父元素就可以了。

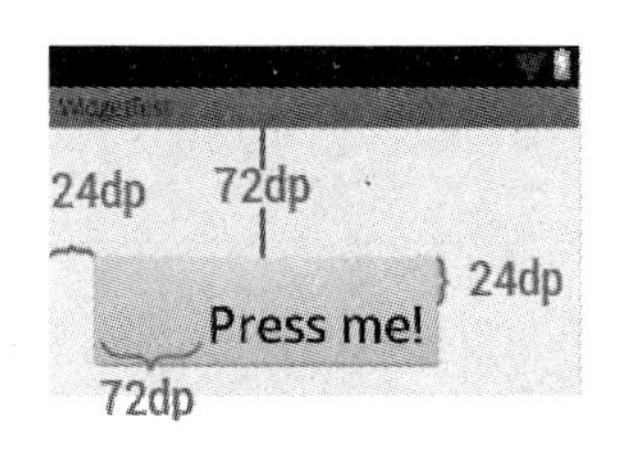

图 2-39　padding 和 margin 的区别

style

当我们需要多个相同格局的控件时,例如所有的 TextView 都是相同的布局:

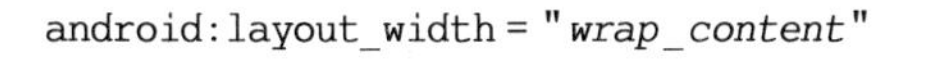

```
android:layout_width = "wrap_content"
```

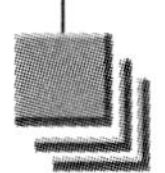

```
android:layout_height = "wrap_content"
```

重复写布局文件就显得很麻烦。我们可以使用 style 来复用相同的代码。可以在 value 文件夹下的 styles. xml 文件中加入代码：

```
<style name = "text_style" >
    <item name = "android:layout_width"> wrap_content </item>
    <item name = "android:layout_height"> wrap_content </item>
    <item name = "android:textSize"> 23dp </item>
</style>
```

在文件中直接使用 style 引用：

```
<TextView
    style = "@style/text_style"
    android:text = "@string/hello_world"/>
```

等同于：

```
<TextView
    android:layout_width = "wrap_content"
    android:layout_height = "wrap_content"
    android:textSize = "23dp"
    android:text = "@string/hello_world"/>
```

文本框和输入框

我们来认识几个不同的文本框及输入框。有这样代码：

```
<!--
textAppearance 设置字体大小
-->
<TextView
    android:id = "@+id/small"
    android:layout_width = "wrap_content"
    android:layout_height = "wrap_content"
    android:text = "@string/small"
    android:textAppearance = "?android:attr/textAppearanceSmall"/>
```

引用了一个 Android 自定义的字体大小，看起来如图 2-40 所示。

Small Text

图 2-40　小字体 TextView

有这样代码：

```
<!--
autoLink 地址可单击
```

```
-->
<TextView
    style="@style/text_style"
    android:id="@+id/url"
    android:autoLink="all"
    android:text="@string/url"/>
```

如果文本内容是一个地址的话，是一个链接的样子，当然也可以是链接，如图 2-41 所示。

http://www.google.com/

图 2-41 链接 TextView

有这样代码：

```
<!--
textStyle  字体样式
background 背景
typeface   字体
-->
<TextView
    android:id="@+id/test"
    style="@style/text_style"
    android:textStyle="italic"
    android:typeface="serif"
    android:background="@color/blur"
    android:text="@string/mutiple"/>
```

这个比较复杂，设置了字体、样式以及背景，看起来如图 2-42 所示。

line 1
line 2
line 3

图 2-42 字体、样式和背景

有这样代码：

```
<!--
ellipsize  当文字太多时,在那个地方省略为...
-->
<TextView
    style="@style/text_style"
    android:ellipsize="end"
    android:singleLine="true"
    android:text="@string/hello_world"/>
```

文字太长显示不下的时候，就会在结尾的地方自动变成省略号，如图 2-43 所示。

Hello world! 标题要长长长长...

图 2-43　结尾自动隐藏

还有这样，当然这是一个平凡的输入框：

```
<!--
普通的输入框
-->
<EditText
    android:layout_width="200dp"
    android:layout_height="wrap_content"
    android:inputType="text"/>
```

作为一个优秀的程序员，怎么能满足于平凡的输入框！

```
<!--
hint          没有输入文字时的提示信息
background    自定义背景,可以是 selector
inputType     设置文本类型,此处为密码,输入的字符会自动变成"*"
drawableLeft  设置输入框左边的图片
-->
<EditText
    android:layout_width="wrap_content"
    android:layout_height="wrap_content"
    android:drawableLeft="@drawable/icon_key"
    android:hint="@string/pwd"
    android:background="@drawable/edittext_style_selector"
    android:inputType="textPassword"
    android:ems="10"/>
```

这个 EditText 的 background 属性引用了 drawable 文件夹下的 xml 文件：

```
<?xml version="1.0" encoding="utf-8"?>
<selector
    xmlns:android="http://schemas.android.com/apk/res/android">
    <item android:state_focused="true"
        android:drawable="@drawable/edittext_style_focused"/>
    <item android:drawable="@drawable/edittext_style_normal"/>
</selector>
```

这个 drawable 文件定义了获得输入焦点(android：state_focused="true")和其他情况下的背景图片。当获得输入焦点时如图 2-44 所示。

当没有获得输入焦点时如图 2-45 所示。

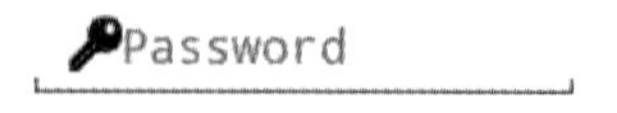

图 2-44　获得焦点的 Edit Text

Password

图 2-45　失去焦点的 Edit Text

按钮

Android 中的按钮有两种：Button 和 ImageButton，顾名思义，前者是普通的按钮，后者是支持图片的按钮。

第一个 Button 是这样：

```
<Button
    android:paddingTop = "4dp"
    android:paddingBottom = "4dp"
    android:paddingLeft = "24dp"

    android:paddingRight = "24dp"
    android:layout_width = "wrap_content"
    android:layout_height = "wrap_content"
    android:background = "@drawable/button_style"
    android:text = "@string/button"
    android:textSize = "30dp"
    android:textColor = "@android:color/white"
    android:onClick = "clickMethod"/>
```

其中 button_style 的定义为：

```
<?xml version = "1.0" encoding = "utf - 8"?>
<selector  xmlns:android = "http://schemas.android.com/apk/res/android">
    <item android:drawable = "@color/yellow" android:state_pressed = "true"></item>
    <item android:drawable = "@color/blue" ></item>
</selector>
```

同输入框一样，当我们按下按钮时如图 2-46 所示。放开按钮时如图 2-47 所示。

Press me!

图 2-46　按下按钮

Press me!

图 2-47　松开按钮

此外，最新的 ADT 支持在 xml 中定义鼠标单击事件的方式：

```
android:onClick = "clickMethod"
```

在使用该 layout 的 Activity 中定义这样一个函数(注意参数为一个 View，且必须为 public)：

```
public void clickMethod(View v){
    Log.e(TAG,"真呀真方便");
}
```

当单击该按钮时，就会调用这个方法。在以前我们需要在 Activity 中找到这个 Button，然后再设置监听器：

```
Button button = (Button) findViewById(R.id.small);
button.setOnClickListener(new OnClickListener(){
    public void onClick(View v) {
    }
});
```

我们使用 ImageButton 来展示一张按钮图片：

```
<ImageButton
    android:layout_width = "wrap_content"
    android:layout_height = "wrap_content"
    android:background = "@drawable/bcg_button"/>
```

如图 2-48 所示。而当我们把 android:background 替换为 android:src 时，悲剧就发生了，如图 2-49 所示。

图 2-48　ImageButton

图 2-49　设置为 src 的 ImageButton

事实上，Button 本身就支持图片作为 background 的行为：

```
<Button
    android:layout_width = "wrap_content"
    android:layout_height = "wrap_content"
    android:background = "@drawable/bcg_button"/>
```

这个效果看起跟第一个一模一样。

读者可能会问，那 ImageButton 有什么意义呢？意义在于，有时候我们在使用非矩形按钮时（如圆形），ImageButton 既可以设置按钮为 Image Source，又可以设置 Background 作为背景（一般为颜色）。这一点是 Button 无法做到的。关于这个问题我们会在以后的章节中讲解。

图片框

顾名思义，ImageView 就是用来展示图片的，最基本的用法是这样：

```
<ImageView
        android:layout_width = "wrap_content"
        android:layout_height = "wrap_content"
        android:src = "@drawable/bcg_button"/>
```

这样看起来十分和谐，如图 2-50 所示。

如果想要在一定的区域内显示你的图片，然后进行适应的缩放，那么要使用下面这个方法：

图 2-50　ImageView

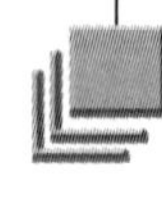

```
android:scaleType
```

我们先定义一个 style，为了便于观察，把这个 50×50 的区域背景设置为黄色：

```
<style name = "img_style">
        <item name = "android:layout_width"> 50dp </item>
        <item name = "android:layout_height"> 50dp </item>
        <item name = "android:background">@color/yellow </item>
        <item name = "android:src">@drawable/bcg_button </item>
</style>
```

有这样：

```
<ImageView
        style = "@style/img_style"
        android:scaleType = "matrix"/>
```

这时图片按照原比例放置，如图 2-51 所示。

图 2-51　scaleType 为 matrix 的 ImageView

如果想让图片填充整个区域，可以这样，如图 2-52 所示。

```
<ImageView
        style = "@style/img_style"
        android:scaleType = "fitXY"/>
```

图 2-52　scaleType 为 fitXY 的 ImageView

或者是让图片自适应控件的大小进行比例缩放，如图 2-53 所示。

```
<ImageView
        style = "@style/img_style"
        android:scaleType = "fitCenter"/>
```

图 2-53　scaleType 为 fitCenter 的 ImageView

同理，fitStart 和 fitEnd 的效果就是把图片的位置放在靠上方或者靠下方，如图 2-54 所示。

图 2-54　fitStart 和 fitEnd 的显示效果

还有这样，如图 2-55 所示。

```
<ImageView
        style="@style/img_style"
        android:scaleType="centerInside"/>
```

图 2-55　scaleType 为 centerInside 的 ImageView

等一下，为什么跟 fitCenter 一模一样？其实是刚好一样而已，但是 centerInside 是有原则的，它只缩不放，从而保证图片的精度。当我把图片缩小 50%时，fitCenter 会把图片放大至适应 ImageView，而 centerInside 会保持图片的原大小，如图 2-56 所示。

图 2-56　缩小显示图片后的结果

还有这样，保持图片原大小居中，如图 2-57 所示。

```
<ImageView
        style="@style/img_style"
        android:scaleType="center"/>
```

图 2-57　scaleType 为 center 的 ImageView

以及这样，填充整个 ImageView，如图 2-58 所示。

```
<ImageView
        style="@style/img_style"
        android:scaleType="centerCrop"/>
```

图 2-58　scaleType 为 centerCrop 的 ImageView

选项控件

有三种基本的选择控件：单选框 RadioGroup、多选框 CheckBox 以及开关 ToggleButton。

RadioGroup

当用 ADT 把 RadioGroup 拖出来时，它如图 2-59 所示。

图 2-59 默认的 RadioButton

```
<RadioGroup
        android:id="@+id/radioGroup2"
        android:layout_width="wrap_content"
        android:layout_height="wrap_content">

        <RadioButton
            android:id="@+id/radio0"
            android:layout_width="wrap_content"
            android:layout_height="wrap_content"
            android:checked="true"
            android:text="RadioButton"/>

        <RadioButton
            android:id="@+id/radio1"
            android:layout_width="wrap_content"
            android:layout_height="wrap_content"
            android:text="RadioButton"/>

        <RadioButton
            android:id="@+id/radio2"
            android:layout_width="wrap_content"
            android:layout_height="wrap_content"
            android:text="RadioButton"/>
</RadioGroup>
```

当然，这个样子太过普通，我们可以根据自己的需求，改变样式。

我们先定义一个样式 radio_style.xml，放在 drawable 文件夹下：

```
<?xml version="1.0" encoding="utf-8"?>
<selector  xmlns:android="http://schemas.android.com/apk/res/android">
    <item android:drawable="@drawable/radio_checked"
```

```
        android:state_checked = "true"></item>
    < item android:drawable = "@drawable/radio_checked1"
        android:state_selected = "true"></item>
    < item android:drawable = "@drawable/radio_checked1"
        android:state_pressed = "true"></item>
    < item android:drawable = "@drawable/checkbox_unchecked1" ></item>
</selector>
```

使用的图片如图 2-60 所示。

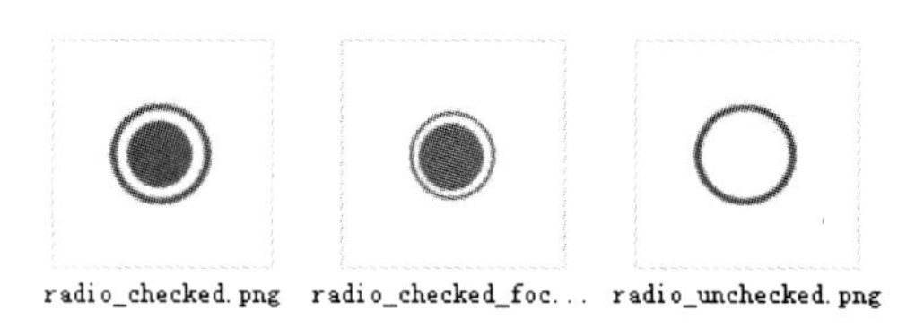

图 2-60　RadioButton 使用的图片

在布局文件中的 RadioButton 中加入这一句：

```
android:button = "@drawable/radio_style"
```

当第一个 RadioButton 已经被选择(checked)，单击第二个 RadioButton(selected)时，它的效果如图 2-61 所示。

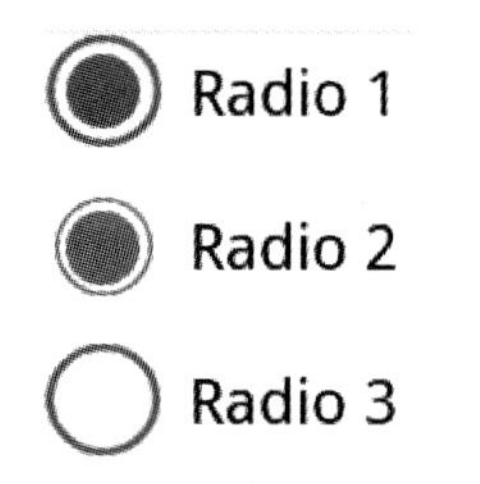

图 2-61　RadioButton 的效果

除此之外，我们还可以对文本的颜色进行控制，在 res 文件夹下新建一个文件夹 color，放入文件 color_radio_group. xml：

```
<?xml version = "1.0" encoding = "utf - 8"?>
< selector xmlns:android = "http://schemas.android.com/apk/res/android">
    < item android:state_checked = "true" android:color = "@color/dark"/>
    < item android:state_selected = "true" android:color = "@color/gray"/>
    < item android:state_pressed = "true" android:color = "@color/gray"/>
    < item android:color = "@color/blur"/>
</selector>
```

在每个 RadioButton 中加入，如图 2-62 所示。

```
android:textColor = "@color/color_radio_group"
```

图 2-62 添加颜色效果的 RadioButton

CheckBox

checkBox 的原生效果如图 2-63 所示。

CheckBox

图 2-63 默认的 CheckBox

我们使用相同的方法对 CheckBox 进行优化，在 CheckBox 的节点里加入：

```
android:button = "@drawable/checkbox_style"
```

checkbox_style 文件的定义如下，放在 drawable 文件夹。

```
<?xml version = "1.0" encoding = "utf - 8"?>
<selector xmlns:android = "http://schemas.android.com/apk/res/android">
    <item android:drawable = "@drawable/checkbox_checked"
            android:state_pressed = "false"
            android:state_checked = "true"></item>
    <item android:drawable = "@drawable/checkbox_checked_focus"
            android:state_pressed = "true"
            android:state_checked = "true"></item>
    <item android:drawable = "@drawable/checkbox_unchecked_focus"
            android:state_pressed = "true"
            android:state_checked = "false"></item>
    <item android:drawable = "@drawable/checkbox_unchecked"
            android:state_pressed = "false"
            android:state_checked = "false"></item>
</selector>
```

使用的图片如图 2-64 所示。

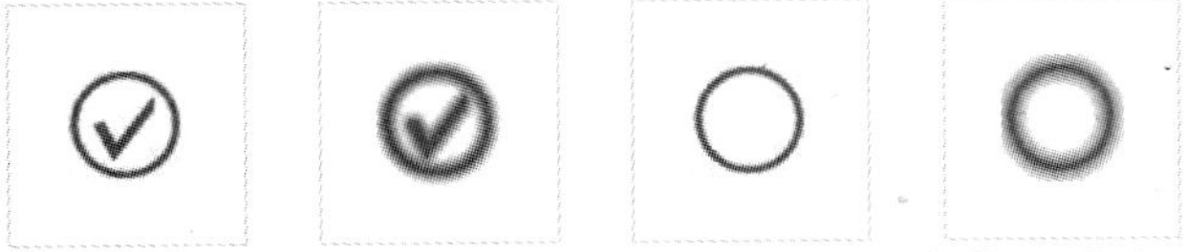

图 2-64 checkbox_style 文件定义的图片

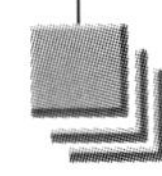

ToggleButton

ToggleButton 的原生效果如图 2-65 所示。

图 2-65　ToggleButton

ToggleButton 本质上是一个 Button，所以我们可以对它进行和 Button 同样的设置：

```
<ToggleButton
        android:paddingTop = "2dp"
        android:paddingBottom = "2dp"
        android:textSize = "20dp"
        android:layout_width = "90dp"
        android:layout_height = "wrap_content"
        android:background = "@color/blue"
        android:textColor = "@android:color/white"
        android:textOn = "@string/text_on"
        android:textOff = "@string/text_off"/>
```

我们重新设置了开关的文字：

```
android:textOn = "@string/text_on"
android:textOff = "@string/text_off"/>
```

因而看到的效果是(左边为 off，右边为 on)如图 2-66 所示。

Female　Male

图 2-66　重设的 ToggleButton

第三章　上课了

打造 Metro 风格界面

清单

Demo 代码：

\demo\Demo_Timetable

实例代码：

\source_codes\Timetable

Target SDK：

Android 2.1

从这一章开始，我们开始全面进入产品开发的节奏。

第一节　产品介绍

两年多前，我刚学习 Android 不久就做了一个课程提醒的应用。当时只是想做一个小东西自己用用，技术不成熟，做得也比较简陋。当时并没有意识到这样的小玩意就是一个大的市场，要知道中国可是有 3000 万大学生。而这样的软件业符合了创业的一个特性：小而精。现在做得比较好的课程表软件有课程格子和超级课程表，如图 3-1 所示。

这一节我们就来自己动手做一个课程表的应用。但是当然，我们最主要还是学习。

需求分析

我们的课表应用有以下几个功能：

- 查看课表；
- 编辑课表；
- 编辑上课时间；
- 上课提醒；
- 桌面插件。

界面设计

课表属于一个三维的结构，星期为一维，时间为一维，上课内容（时间、地点等）为另一

图 3-1　课程格子应用

维，所以使用列表的方式来展现数据。我们借鉴 Summly 的设计风格，使用大色块的列表，并使用模糊的效果作为背景，如图 3-2 所示。

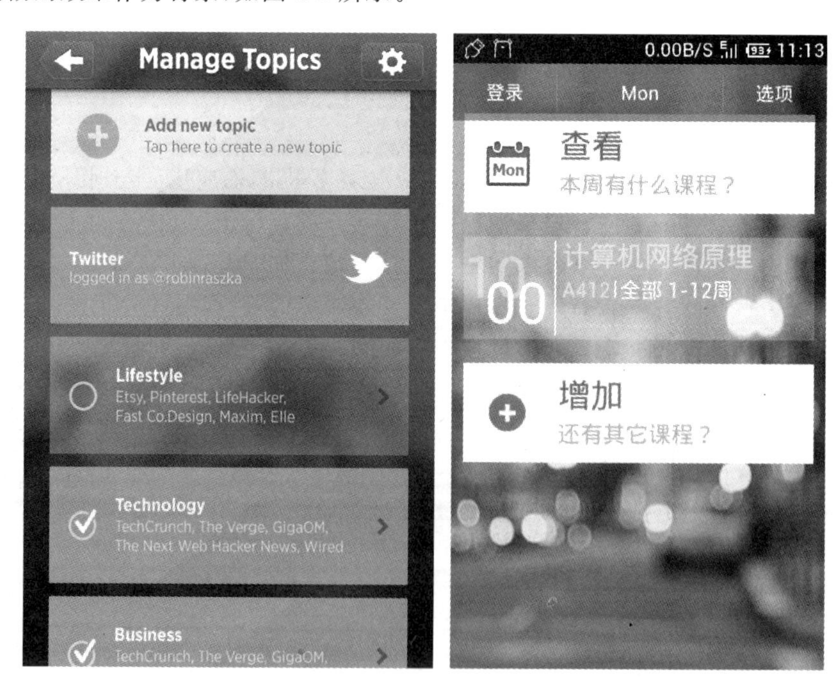

图 3-2　Summly(左)与本书实例应用(右)

在设置页面上，延续了模糊的背景，但是列表的展示换成了二级列表，如图 3-3 所示。

图 3-3　设置界面

用户体验设计

在这个应用里面，我们使用了自定义的时间选择和日期选择的控件。

如图 3-4 所示，在时间选择一栏中用户可能会感到困惑，因为按照一贯的逻辑，要修改时间的操作应该是“单击”→“输入”。所以当用户单击这个框的时候会发现没有达到预期的效果，甚至放弃操作。

于是我进行了这样的一个设计：在界面一打开的时候播放一个滚动的动画，提示用户这个地方是要上下翻动选择的，这样既节省了提示的空间，又能让用户觉得新奇，如图 3-5 所示。

图 3-4　添加课程界面

图 3-5　时间栏滚动显示

此外，因为有些课程是从第某周开始，到第某周结束，因而我设计了一个勾选的动画：当勾选“选择”时，选择的内容从右边滑入；而取消选择时，又会向右边滑出，如图 3-6 所示。

再来看这个“概念时钟”，如图 3-7 所示。你可以把它当作一个反例，因为它确实是一个反例。它的缺点有一大堆。其一，如果没有我说明这是一个选择时钟的控件，大概没有人会想到。这一点严重违反了“don't make me think”的设计信条；其二，用户很难选择出精确

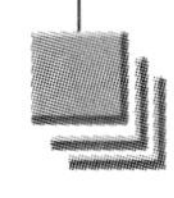

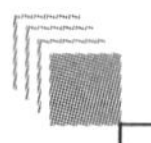

到每一分钟的时间。

图 3-6　滑动效果　　　　图 3-7　时间选择控件

时间选择的控件涉及到 SurfaceView 的知识，因而会把这个控件的代码放到下面一章中去讲解。

第二节　数据展示 ListView

ListView 也就是列表，列表在 Android 中有很多的使用，例如 Android 的系统设置界面，如图 3-8 所示。

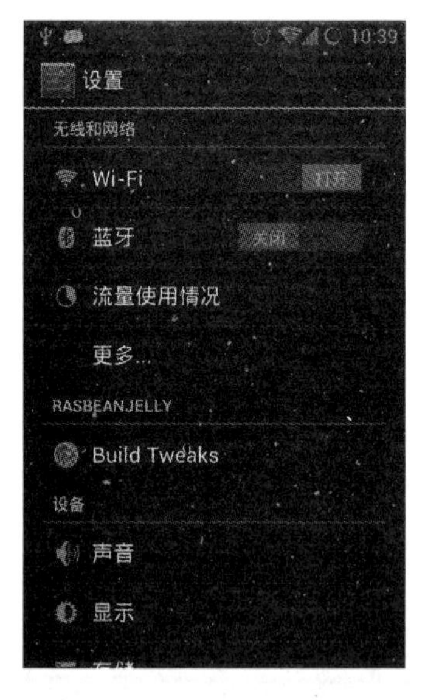

图 3-8　Android 设置界面

列表的显示需要三个组件：

- ListView　列表。
- Adapter　用来把数据放到 ListView 上的适配器。
- 数据　可以是数据、图片。

其中 Adapter 有四种，它们分别是 BaseAdapter、ArrayAdapter、SimpleAdapter 和 SimpleCursorAdapter。

其中 ArrayAdapter 最简单，只能展示一行字；BaseAdapter 是一个抽象类，所以灵活性最强；SimpleAdapter 扩展性最好，可以自定义出各种效果；SimpleCursorAdapter 是 Cursor 的一种方便的使用方法，可以把数据库的内容以列表的形式展示出来。

我们在这一章中要使用的是最复杂的 BaseAdapter。

配置 ListView 布局

我们在布局文件中定义一个 ListView：

```
<ListView
    android:id="@+id/listView"
    android:layout_width="fill_parent"
    android:layout_height="wrap_content">
</ListView>
```

这个 ListView 只是告诉父容器这个布局中有一个列表视图，但是列表视图里面的内容是什么样的，我们要继续定义。新建一个文件 layout_listview.xml：

```
<?xml version="1.0" encoding="utf-8"?>
<LinearLayout xmlns:android="http://schemas.android.com/apk/res/android"
    android:layout_width="fill_parent"
    android:layout_height="fill_parent"
    android:orientation="horizontal" >

    <ImageView
        android:id="@+id/listview_head_iv"
        android:layout_width="60dp"

        android:layout_height="60dp"
        android:src="@drawable/ic_launcher"/>

    <TextView
         android:layout_width="8dp"
         android:layout_height="wrap_content"/>

    <LinearLayout
        android:layout_width="fill_parent"
        android:layout_height="60dp"
        android:gravity="center_vertical"
        android:orientation="vertical" >

            <TextView
            android:id="@+id/listview_title_tv"
            android:layout_width="wrap_content"
            android:layout_height="wrap_content"
            android:text="@string/temp_title"
            android:textSize="18dp"/>
        <TextView
            android:id="@+id/listview_num_tv"
```

```
            android:layout_width = "wrap_content"
            android:layout_height = "wrap_content"
            android:textColor = "#777777"
            android:textSize = "15dp"
            android:text = "@string/temp_num"/>
    </LinearLayout>

</LinearLayout>
```

它在视图中看起来如图 3-9 所示。

图 3-9 视图显示

我们继续在 Java 代码中加入适配器,把自己需要的数据写入到列表中。首先新建一个实体类:

```
public class Contacts {

    //联系人姓名
    private String name;
    //联系人号码
    private String tel;
    //联系人头像资源 id
    private int head;
    //省略 get set 以及构造函数方法
```

新建一个 Adapter 类,这个 Adapter 决定了这个 ListView 有几个元素(getCount()),每个元素的视图是怎样的(getView)。

```
public class MyAdapter extends BaseAdapter{

    private ArrayList<Contacts> list;
    private Context context;

    public MyAdapter(Context context,ArrayList<Contacts> list){
        this.context = context;
        this.list = list;
    }

    @Override
    public int getCount() {
        //返回列表长度
        return list.size();
    }

    @Override
    public View getView(int index,View view,ViewGroup arg2) {
```

```
        //返回列表中第 index 个视图
        Contacts c = list.get(index);
        if(view == null){
            //设置布局
            view = LayoutInflater.from(context).inflate(
                    R.layout.layout_listview,null);
        }
        //查找控件
        ImageView headIv = (ImageView) view.findViewById(R.id.listview_head_iv);
        TextView nameTv = (TextView) view.findViewById(R.id.listview_title_tv);
        TextView numTv = (TextView) view.findViewById(R.id.listview_num_tv);
        //设置视图内容
        headIv.setImageResource(c.getHead());
        nameTv.setText(c.getName());
        numTv.setText(c.getTel());
        return view;
    }

    @Override
    public Object getItem(int index) {
        return list.get(index);
    }
    @Override
    public long getItemId(int index) {
        return index;
    }
```

在 Activity 中添加这个适配器：

```
@Override
public void onCreate(Bundle savedInstanceState) {
    super.onCreate(savedInstanceState);
    setContentView(R.layout.activity_main);

    ListView listview = (ListView) findViewById(R.id.listView);
    //设置适配器
    listview.setAdapter(new MyAdapter(this,getDatas()));
}

/**
 * 模拟获取数据
 * @return
 */

private ArrayList<Contacts> getDatas(){
    ArrayList<Contacts> list = new ArrayList<Contacts>();
    list.add(new Contacts("Shinado","13333333333",R.drawable.head));
    list.add(new Contacts("Neal","15333333333",R.drawable.ic_launcher));
    return list;
}
```

结果如图 3-10 所示。

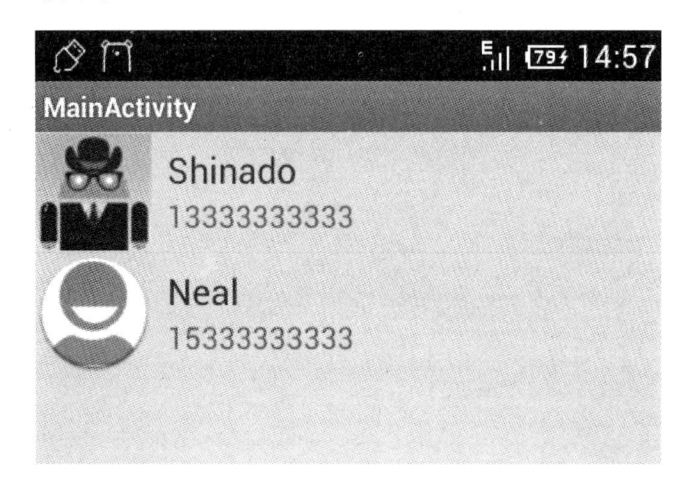

图 3-10　列表显示

ListView 详解

ListView 的工作原理如下：

- ListView 针对 List 中每个 item，要求 adapter“给我一个视图”(getView)。
- 一个新的视图被返回并显示。

我们把 getView 方法改成这样：

```
@Override
public View getView(int index,View view,ViewGroup arg2) {
     Log.e(TAG,"getView():" + index + " null? " + (view == null));
    //返回列表中第 index 个视图
    Contacts c = list.get(index);
    if(view == null){
        //设置布局
        view = LayoutInflater.from(context).inflate(
                    R.layout.layout_listview,null);
        //查找控件
        ImageView headIv = (ImageView) view.findViewById(
                    R.id.listview_head_iv);
        TextView nameTv = (TextView) view.findViewById(
                    R.id.listview_title_tv);
        TextView numTv = (TextView) view.findViewById(R.id.listview_num_tv);
        //设置视图内容
        headIv.setImageResource(c.getHead());
        nameTv.setText(c.getName());
        numTv.setText(c.getTel());
    }
    return view;
}
```

这样看起来似乎很合理，每次系统调用 getView 方法时，只有在 view 为空的情况下才

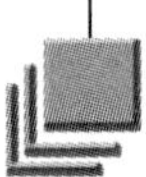

需要设置视图内容，不为空时直接返回。运行一下，嗯，没有出什么错，如图 3-11 所示。

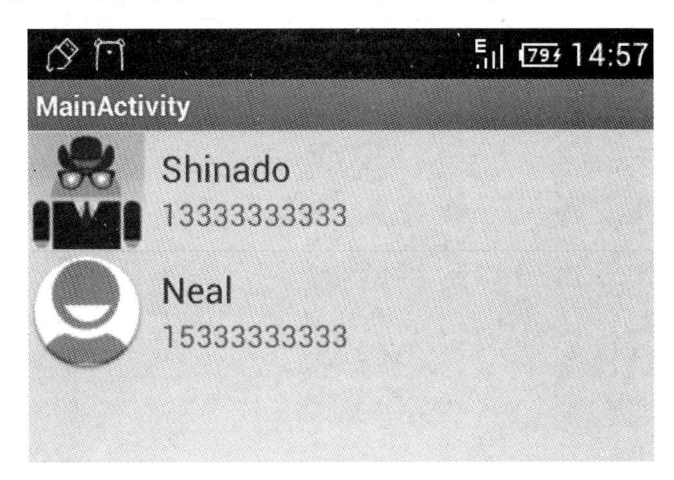

图 3-11　只有两列的列表

但是当有 10 条数据的时候，悲剧就发生了：向下滚动时，数据乱套了，如图 3-12 所示。

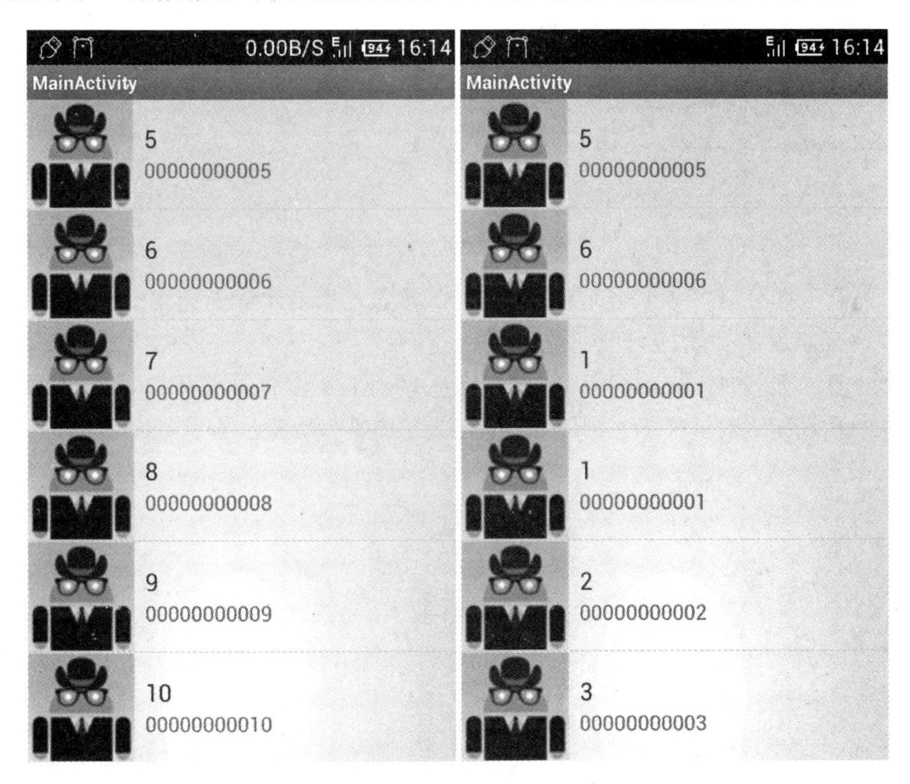

图 3-12　实际效果（右图）没有达到预期

这是因为 ListView 使用了缓存的机制，它的原理大概是这样：

- 不管你有多少个数据，其中只有可见的数据存在内存中，其他的在 Recycler 中。
- ListView 先请求一个视图（getView），然后请求其他可见的视图。convertView 在 getView 中是空（null）的。

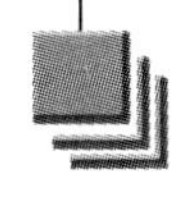

- 当 item1 滚出屏幕，并且一个新的项目从屏幕底端上来时，ListView 再请求一个视图。convertView 此时不是空值了，它的值是 item1。你只需设定新的数据然后返回 convertView，不必重新创建一个视图，如图 3-13 所示。

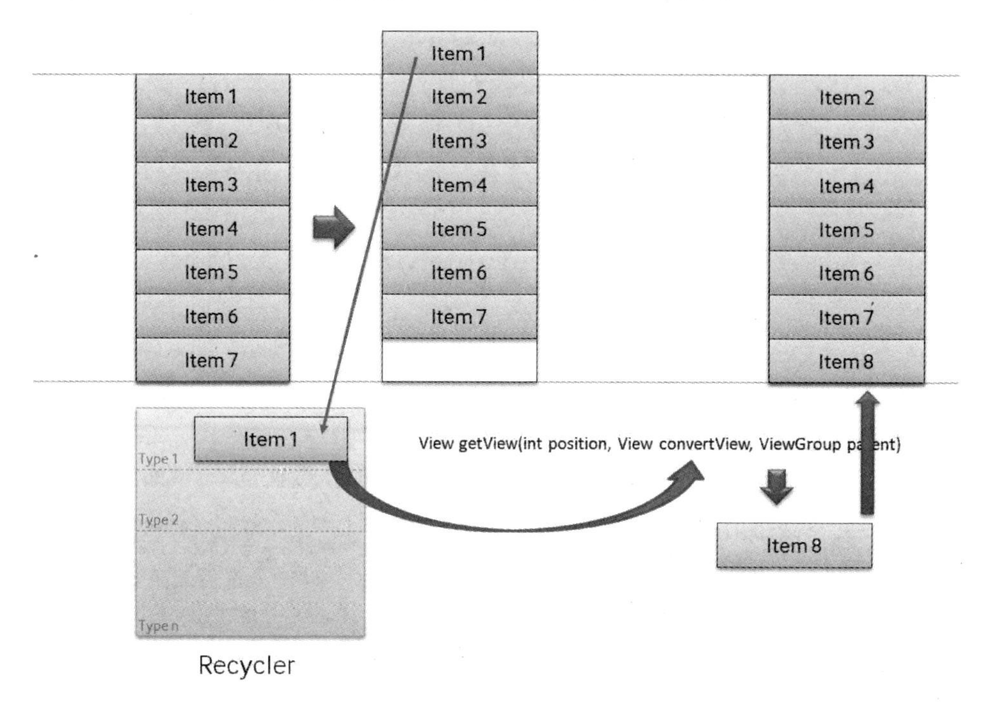

图 3-13　ListView 视图加载示意

因而屏幕上能放得下多少个元素，ListView 就有多少个实际的元素。你翻，或者不翻，ListView 的元素就那么多，不增不减。

再来分析一下 getView 的代码，每次调用 getView 方法都会调用 findViewById。既然不管是哪个元素它的布局都是相同的，那么就没有必要每次都去查找布局了。因此可以改成这样：

```
@Override
public View getView(final int index,View view,ViewGroup arg2) {
    Log.e(TAG,"getView():" + index + " null? " + (view == null));
    //返回列表中第 index 个视图
    Contacts c = list.get(index);
    ViewHolder holder = null;
    if(view == null){
        //设置布局
        view = LayoutInflater.from(context).inflate(
                    R.layout.layout_listview,null);
        //查找控件
        holder = new ViewHolder();
        holder.headIv = (ImageView) view.findViewById(R.id.listview_head_iv);
        holder.nameTv = (TextView) view.findViewById(R.id.listview_title_tv);
        holder.numTv = (TextView) view.findViewById(R.id.listview_num_tv);
```

```
            //绑定
            view.setTag(holder);
        }else{
            //获取绑定的 holder
            holder = (ViewHolder) view.getTag();
        }
        //设置视图内容
        holder.headIv.setImageResource(c.getHead());
        holder.nameTv.setText(c.getName());
        holder.numTv.setText(c.getTel());
        return view;
    }
    static class ViewHolder{
        ImageView headIv;
        TextView nameTv;
        TextView numTv;
    }
```

如果不熟悉 Adapter 的话，它会让你抓狂；如果你抓住要点，它是不会逃出你的“五指山”的。记住一点：一旦 ListView 加载完成，getView 方法里面的 view 就不会为空。

第三节　数据存储

Android 提供了五种存储方法，它们分别为 SharedPreferences、文件存储、SQLite 存储、ContentProvider、网络存储。其中 SharedPreferences 为轻量级的“键-值”存储；文件存储是利用 I/O 流来读取文件的；SQLite 就是数据库存储；ContentProvider 是一个特殊的存储数据的类型，它提供了标准的接口用来获取、操作数据；网络存储顾名思义就是把数据存放在网络的服务器中。

我们在本节中将会学习 SQLite 和 SharedPreferences。在 Android 上使用数据库有一点不好的是没有可视化界面，创建数据库、创建表都需要用代码来实现。并且我们需要额外的工具来查看数据库中的数据。

查看应用的数据库需要获取 Root 权限，然后使用第三方应用来查看。获取 Root 权限不是我们的重点，读者可以上网查阅教程。

使用 SQLite 存储数据

要使用 Android 的数据库服务，需要继承 SQLiteOpenHelper 这个类。我们一般使用这样的架构来搭建一个数据库：

```
public class DBUtil extends SQLiteOpenHelper{

    public DBUtil(Context context) {
        //创建数据库
        super(context,DBValue.DB_NAME,null,1);
```

```
    }

    @Override
    public void onCreate(SQLiteDatabase db) {
        //建表,可能有多个表
        db.execSQL(DBValue.Table_Time.CREATE_TABLE);

        //插入初始数据
        initTimes(db);
    }
     ......
```

onCreate 这个函数在数据库第一次生成(不是每次 new 一个对象时)的时候就会调用,也就是说当程序第一次打开的时候调用这个方法,因此我们在这个方法里面生成表以及初始化数据。

DBValue 用来保存数据库的名称、表名以及字段等静态信息:

```
/**
 * 结构为{数据库{表{字段}}}
 * @author Administrator
 *
 */
public class DBValue {

    //数据库名
    public static final String DB_NAME = "timetable.db";
    //自增主键
    private static final String AUTO_INCREMENT = "INTEGER PRIMARY KEY AUTOINCREMENT";

    //time表
    public static class Table_Time {
        //表名
        public static final String TABLE_NAME = "time";

        //字段
        public static final String TIME_ID_COL = "time_id";
        public static final String TIME_COL = "time";
        public static final String[] TABLE_COL = new String [] {
            TIME_ID_COL,
            TIME_COL
        };

        public static final String SELECTORDER = "time_id DESC";

        //创建表代码
        public static final String CREATE_TABLE = "CREATE TABLE IF NOT EXISTS " + TABLE_NAME + "(" +
                TIME_ID_COL + " " + AUTO_INCREMENT + "," +
                TIME_COL + " VARCHAR(10))";
    }
}
```

Android 提供了两种方法来对数据库进行操作，一种是 SQL 语句，另一种是使用由 SQLiteDatabase 封装的方法。我们先通过 SQLiteOpenHelper. getWritableDatabase()或者 getReadeableDatabase()方法来获取 SQLiteDatabase 对象。二者的区别也显而易见，前一个是获取写入(增删改)，后一个是获取读出(查)的 SQLiteDatabase。

我们先来看数据库查询的方法。Android 查询数据库的方法为：

```
Cursor query(String table,String[] columns,String selection,
    String[] selectionArgs,String groupBy,String having,String orderBy)
```

参数解释如下。

- table：要查询的表名。
- columns：一个投影，也就是你需要返回的列名。
- selection：一个 where 子句定义了要返回的行，包含"?"。
- selectionArgs：一个选择参数字符串的数组，它将会替换 where 子句中的"?"。
- groupBy：一个 group by 子句，用来定义返回的行的分组方式。
- having：一个 having 过滤器。
- orderBy：要返回的行的顺序。

例如：

```
db = dbUtil.getReadableDatabase();

        String whereClause = "time_id = ?";
        String[] whereArgs = {"1"};
        Cursor c = db.query("time",new String[]{"time_id","time_value"},
                whereClause,whereArgs,null,null,"time_id DESC");
```

这段代码换成 SQL 语句就是：

```
select time_id,time_value from time where time_id = 1
```

然后对返回的 Cursor 进行遍历操作：

```
while(c.moveToNext()){
    int index_id = c.getColumnIndex("time_id");
    int id = c.getInt(index_id);
    int index_value = c.getColumnIndex("time_value");
    String value = c.getString(index_value);
}
```

更新以及插入操作的方法都需要使用 ContentValues 这个对象，它相当于一个键值对，用来设置字段的值。先来看插入方法：

```
public void testInsert() {
    db = dbUtil.getWritableDatabase();

    ContentValues cv = new ContentValues();
```

```
    cv.put("time_value","10:10");

    db.insert("time",null,cv);
    db.close();
}
```

更新数据表的方法：

```
public void testUpdate(){
    db = dbUtil.getWritableDatabase();

    ContentValues cv = new ContentValues();
    cv.put("time_value","10:20");

    String whereClause = "time_value = ?";
    String[] whereArgs = {"10:10"};
    db.update("time",cv,whereClause,whereArgs);
    db.close();
}
```

删除表：

```
public void testDelete() {
    db = dbUtil.getWritableDatabase();

    String whereClause = "time_value = ?";
    String[] whereArgs = {"10:20"};
    db.delete("time",whereClause,whereArgs);
    db.close();
}
```

使用SharedPreferences存储数据

SharedPreferences 是一个轻量级的键值对存储方式，它的本质是 XML 文件。SharedPreferences 的使用非常简单：

```
/*
 * 获取 SharedPreference
 * "test"就是 SharedPreference 的名,像表名一样
 */
SharedPreferences preferences = getSharedPreferences("test",MODE_PRIVATE);
//获取 Editor 对象用于写入
Editor editor = preferences.edit();
editor.putString("name","Shinado");
//提交修改
editor.commit();

//获取值
String name = preferences.getString("name","");
```

第四节　通知 Notification

延迟的意图 PendingIntent

我们知道，Intent 是一个桥梁的作用，而我们接下来要讲的 PendingIntent 跟 Intent 类似，但不同的是通过 Intent 来开始服务、Activity 或者其他功能的时候是马上执行的，而 PendingIntent 不会马上执行，而是等系统处理、或者用户做了某个动作之后再执行的。也就是说 PendingIntent 就是一个延迟发送的 Intent。

要得到一个 PendingIntent 对象，需要使用该类的静态方法 getActivity(Context,int,Intent,int)、getBroadcast(Context,int,Intent,int)和 getService(Context,int,Intent,int)。它们分别对应着 Intent 的三个行为，跳转到一个 Activity、打开一个广播和打开一个服务。

一般来说，我们只需要关注第一个参数 Context 和第三个参数 Intent，第二个参数和第四个参数一般为 0。第一个参数不多说了，大家也很熟悉了；第三个参数就是我们定义的行为。

在本节我们使用 PendingIntent 来在通知栏中打开 Activity；在下一节我们也需要使用 PendingIntent 来实现插件的事件处理。

为了便于读者理解，我新建了一个工程用于测试 PendingIntent 的功能。这个工程很简单，就一个 Activity 加一个 BroadcastReceiver，Activity 负责发送短信，并且通过 PendingIntent 来发送广播。

```
/*
 * PendingIntent 测试
 */
String msg = "发短信要钱吗?";
String number = "10086";
SmsManager sms = SmsManager.getDefault();

String action_sent = "com.shinado.sms.sent";
String action_delivery = "com.shinado.sms.delivery";
PendingIntent sent = PendingIntent.getBroadcast(this,0,
        new Intent(action_sent),0);
PendingIntent delivery = PendingIntent.getBroadcast(this,0,
        new Intent(action_delivery),0);
/*
 * sent: 信息发送的广播
 * delivery: 信息送达用户的广播
 */
sms.sendTextMessage(number,null,msg,sent,delivery);
Log.e(TAG,"right now");

//注册广播接收器
IntentFilter filter = new IntentFilter();
filter.addAction(action_sent);
filter.addAction(action_delivery);
registerReceiver(receiver,filter);
```

广播接收器的代码为：

```
BroadcastReceiver receiver = new BroadcastReceiver(){
    @Override
    public void onReceive(Context context, Intent intent) {
        Log.e(TAG, intent.getAction());
    }
};
```

特别注意

发送短信需要权限，在 Manifest.xml 的根节点中加入：

```
<uses-permission android:name="android.permission.SEND_SMS"/>
```

查看 Logcat 的输出，如图 3-14 所示。

Time	Text
04-14 11:09:32.492	right now
04-14 11:09:35.152	com.shinado.sms.sent
04-14 11:09:35.629	com.shinado.sms.delivery

图 3-14　AppWidgetTest 运行结果

请大家注意 Log 的时间，大家可以看到，第一条 Log 信息（也就是在 Activity 中发送完短信之后）的时间为 34 分 45 秒，而广播接收到的发送短信时间为 35 分 04 秒。这 9 秒就是用户按下“发送”和系统发送短信的时间。也就是说，使用 PendingIntent 不是马上发送 Broadcast，而是等系统真正发完短信之后才发送 Broadcast 的。

创建通知

要在通知栏中创建一个通知，我们需要使用两个类：NotificationManager 和 Notification。NotificationManager 是通知栏的管理器，它有三个公共方法：

- cancel(int id)　取消以前显示的一个通知。
- cancelAll()　取消以前显示的所有通知。
- notify(int id, Notification notification)　显示通知。

获取 NotificationManager 的方法为：

```
myNotiManager = (NotificationManager)getSystemService(NOTIFICATION_SERVICE);
```

Notification 就是通知的一个实例，它可以设置的属性及解释如表 3-1 所示。

表 3-1　Notification 属性一览

公有属性	类　型	解　释
audioStreamType	int	当声音响起时，所用的音频流的类型
contentIntent	PendingIntent	当通知条目被单击，就执行这个被设置的 PendtingIntent
contentView	RemoteViews	当通知被显示在状态条上的时候，同时这个被设置的视图被显示
defaults	int	默认值
deleteIntent	PendingIntent	当用户单击“清除”按钮删除所有的通知时所执行的 PendingIntent

续表

公有属性	类　型	解　释
flags	int	不解释(什么？API都没有解释?)
fullScreenIntent	PendingIntent	当用户拉出通知栏时所执行的 PendingIntent
icon	int	状态条所用的图片
iconLevel	int	假如状态条的图片有几个级别,就设置这里
largeIcon	Bitmap	可能会超出状态栏的大图片
ledARGB	int	LED灯的颜色
ledOffMS	int	LED关闭时的闪光时间(以毫秒计算)
ledOnMS	int	开始时的闪光时间(以毫秒计算)
number	int	这个通知代表事件的号码
sound	Uri	通知的声音
tickerText	CharSequence	通知被显示在状态条时所显示的信息
tickerView	RemoteViews	通知被显示在状态条时所显示的窗口
vibrate	long[]	振动模式
when	long	通知的时间戳

除此之外,Notification 还有一个重要的方法:

```
setLatestEventInfo(Context context,CharSequence contentTitle,
        CharSequence   contentText,PendingIntent contentIntent)
```

这个方法是创建 Notification 时必不可少的。它的参数解释如下:

- context　Activity 的引用。
- contentTitle　通知的标题。
- contentText　通知的内容。
- contentIntent　将会执行的内容(一般是打开一个 Activity)。

我们测试一个例子:

```
/*
 * Notification 测试
 */
NotificationManager myNotiManager =
                (NotificationManager)getSystemService(NOTIFICATION_SERVICE);

//这个 Intent 的意图为打开 NewActivity 这个 activity
Intent notifyIntent = new Intent(this,NewActivity.class);
notifyIntent.setFlags(Intent.FLAG_ACTIVITY_NEW_TASK);

PendingIntent appIntent = PendingIntent.getActivity(this,0,notifyIntent,0);

Notification noti = new Notification();
//图标
noti.icon = R.drawable.head;
//声音
```

```
noti.defaults = Notification.DEFAULT_SOUND;
//单击通知后自动消失
noti.flags = Notification.FLAG_AUTO_CANCEL;
//设置事件
noti.setLatestEventInfo(this,"title","Hello World!",appIntent);
myNotiManager.notify(0,noti);
```

运行效果如图 3-15 所示。

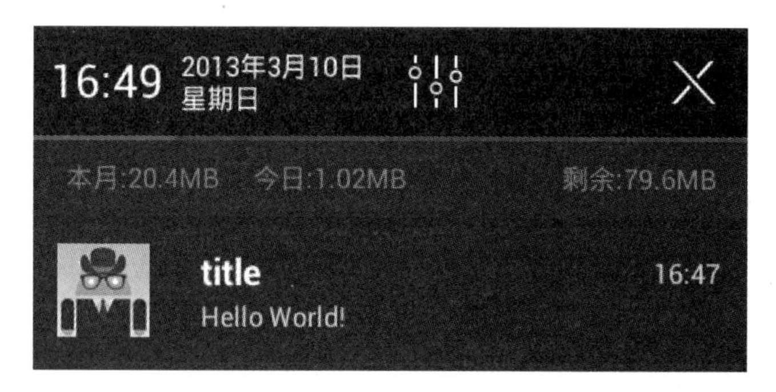

图 3-15 NotificationTest 运行结果

一般来说，通知“寄生”在一个 Service 或者 BroadcastReceiver 之下，在后台中运行，直到特定的情况下才创建通知。短信通知就是一个典型的通知，它使用一个 BroadcastReceiver 在后台中接收短信，一旦收到短信就创建通知。

第五节 桌面插件 AppWidget

开发桌面小插件需要四个步骤：第一，配置 appwidget-provider；第二，配置布局；第三，继承 AppWidgetProvider；第四，在 AndroidMenifest.xml 中添加 receiver。桌面小插件的主要工作都是由 AppWidgetProvider 这个类来完成的，它是 BroadcastReceiver 的一个子类，通过接收广播来更新小插件的显示信息。

配置 appwidget-provider 和布局

我们使用 xml 来配置 appwidget-provider，我们新建一个 xml 文件 appwidget_provider.xml。这个文件包含了桌面插件的长宽、更新时间和布局等信息。

```
<?xml version = "1.0" encoding = "utf - 8"?>
<appwidget - provider xmlns:android = "http://schemas.android.com/apk/res/android"
    android:minWidth = "146dp"
    android:minHeight = "146dp"
    android:initialLayout = "@layout/layout_appwidget" >
</appwidget - provider >
```

定义的布局文件 layout_appwidget.xml：

```
<?xml version = "1.0" encoding = "utf - 8"?>
<RelativeLayout xmlns:android = "http://schemas.android.com/apk/res/android"
    android:id = "@+ id/root"
    android:layout_width = "fill_parent"
    android:layout_height = "fill_parent"
    android:orientation = "vertical" >

    <TextView
        android:id = "@+ id/hour_tv"
        android:layout_width = "wrap_content"
        android:layout_height = "wrap_content"
        android:layout_alignParentLeft = "true"
        android:layout_alignParentTop = "true"
        android:textSize = "100dp"
        android:textColor = "@color/white"/>

    <TextView
        android:id = "@+ id/min_tv"
        android:layout_width = "wrap_content"
        android:layout_height = "wrap_content"
        android:layout_alignParentLeft = "true"
        android:layout_alignParentTop = "true"
        android:layout_marginLeft = "25dp"
        android:layout_marginTop = "32dp"
        android:textColor = "@color/background_white"
        android:textSize = "100dp"/>

</RelativeLayout>
```

在黑色背景下看起来如图 3-16 所示。

再回到 appwidget_provider.xml 文件中。顾名思义 android:minHeight 和 android:minWidth 参数跟插件的长宽有关，但是又不同于实际意义上的长跟宽。Google 提供的基本原则是用你想占用的单元格数量乘以 74，再减去 2。我们的小插件长度占 3 个单元格、宽度占 1 个单元格，因此长为 74×3－2＝220，宽为 74×1－2＝72。但事实上，当我把 android:minWidth 这个参数设置为 160dp 到 239dp 之间的时候插件的宽度都是坚定不移的 3 个单元格。

图 3-16　布局文件显示效果

而 android:updatePeriodMillis 属性理论上是设置 Widget 页面的更新页面的时间的频率，但事实上经过多方证实这个参数的设置是没有用的（至少在 SDK 2.2 中是如此），我们需要在 Java 代码中加入定时器来刷新插件。

比较靠谱的是 android:initialLayout 这个属性，它会坚定不移地按照你的想法来设置初始化页面的布局，我们通过它来从 layout 文件中指定初始化布局，布局文件我们已经介绍过，这里不再赘述。

需要注意的是，配置 appwidget-provider 的文件需放在/res/xml 这个文件夹里面，这个

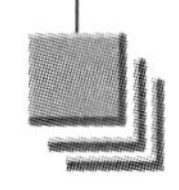

文件夹在工程创建的时候是不存在的，需要我们自己新建，如图 3-17 所示。

图 3-17　新建 xml 文件夹

继承 AppWidgetProvider 和添加 receiver

我们要使用小插件，就必须继承 AppWidgetProvider 这个类。AppWidgetProvider 继承于 BroadcastReceiver，所以它拥有 onReceive 方法，用来接收用户自定义的信息以及 AppWidgetProvider 本身发出的消息。除了 onReceiver 方法外，它还有以下四个方法，构成它的生命周期：

- onUpdate　用户向桌面添加插件、用户删除插件时调用。
- onDelete　当插件被删除时调用。
- onEnable　当第一个插件被创建时调用。
- onDisable　当最后一个插件被删除后调用。

```
public class MyAppWidgetProvider extends AppWidgetProvider{

    @Override
    public void onReceive(Context context, Intent intent){
        /*
         * 用于接收用户定义的广播
         * 比如可以接收触碰事件，然后在这里更新插件
         */
        super.onReceive(context, intent);
    }
    @Override
    public void onEnabled(Context context){
        /*
         * 当桌面上第一个插件被创建时调用
         * 做一些初始化处理，比如创建事件响应、开始服务
         */
        super.onEnabled(context);
    }
    @Override
    public void onUpdate(Context context, AppWidgetManager appWidgetManager, int[] appWidgetIds){
        /*
         * 大杂烩，被删除时、被添加时都会调用
         *
         */
        super.onUpdate(context, appWidgetManager, appWidgetIds);
    }
    @Override
    public void onDeleted(Context context, int[] appWidgetIds){
```

```
        /*
          * 每删除一个插件时都会调用
          *
          */
        super.onDeleted(context,appWidgetIds);
    }
    @Override
    public void onDisabled(Context context){
        /*
          * 删除最后一个插件时会调用
          * 做一些收尾工作,注销服务等
          */
        super.onDisabled(context);
    }
}
```

在这里我稍微讲解一下 AndroidManifest.xml 在 Android 系统中的作用。我们知道，创建一个 Android 应用需要在 AndroidManifest.xml 中声明 activity，并且在第一个启动的 activity 中添加 intent-filter。这个 intent-filter 实际上是给 Android 系统看的。我们知道在启动器中可以看到我们的应用图标，这是因为启动器查看了系统中的所有包含 CATEGORY_LAUNCHER 的 activity。同理，要想让启动器知道我们的桌面小插件，就必须在 AndroidManifest.xml 文件中添加这个 intent-filter。

注意

还需另外加入 android.appwidget.action.APPWIDGET_UPDATE 这个 action，否则当你想要添加你的小插件的时候会惊讶地发现怎么也找不到你的插件了。

数据定时更新和事件响应

我们在之前提到在 appwidget-provider 中定义的定时刷新时间实际上是没有效果的，因此我们需要一个定时器来帮我们自动刷新插件的界面变化。

```
class MyTime extends TimerTask{

        private Context context;
        private AppWidgetManager appWidgetManager;

    public MyTime(Context context,AppWidgetManager appWidgetManager){
        this.context = context;
        this.appWidgetManager = appWidgetManager;
    }
    @Override
    public void run() {
        /*
          * 定时更新 AppWidget
          */
        updateView(context,appWidgetManager);
    }
```

```
}
public void updateView(Context context,AppWidgetManager wManager)
{
    views = new RemoteViews(context.getPackageName(),R.layout.layout_appwidget);
    TimeVo time = new TimeVo();
    views.setTextViewText(R.id.hour_tv,time.getHourString());
    views.setTextViewText(R.id.min_tv,time.getMinString());
    wManager.updateAppWidget(new ComponentName(
        context,MyAppWidgetProvider.class),views);
}
```

然后在 onEnabled 函数中开始定时器、添加事件监听：

```
timer = new Timer();
timer.scheduleAtFixedRate(new MyTime(
        context,AppWidgetManager.getInstance(context)),1,60000);
```

在 onDisabled 中取消定时器：

```
if(timer != null){
    timer.cancel();
}
```

同样我们在 onEnabled 中添加事件响应：

```
//获取布局的 view
views = new RemoteViews(context.getPackageName(),R.layout.layout_appwidget);
AppWidgetManager appWidgetManager = AppWidgetManager.getInstance(context);
ComponentName provider = new ComponentName(context,MyAppWidgetProvider.class);

//更新插件,在这里表现为发送广播
Intent openIntent = new Intent(context,MyAppWidgetProvider.class);
openIntent.setAction(ACTION_UPDATE);
PendingIntent addPendingIntent =   PendingIntent.getBroadcast(
        context,0,openIntent,0);
views.setOnClickPendingIntent(R.id.root,addPendingIntent);
appWidgetManager.updateAppWidget(provider,views);
```

一般来说，Java 以及大部分 Android 的事件响应都是通过实现某个监听器来实现对事件的监听的。但是 AppWidget 的事件监听比较特殊，它是通过发送 Broadcast、接收 Broadcast 来实现事件的。

```
@Override
public void onReceive(Context context,Intent intent){
    Log.e(TAG,"onReceive");
    if(intent.getAction().equals(ACTION_UPDATE)){
        AppWidgetManager am = AppWidgetManager.getInstance(context);
```

```
            updateView(context,am);
        }
        super.onReceive(context,intent);
    }
```

AppWidget 事件实现的途径有两种：一为发送广播，然后重写 onReceiver 来接收；另一为设置 Intent 来打开 Activity。当然，要实现单击插件打开 Activity 也可以使用第一种方法，然后再根据 action 在 onReceiver 方法中打开响应的 Activity。这里就用到了我们前一节所讲到的 PendingIntent 组件，在此就不多加赘述了。

第六节　功能实现

数据库及实体类设计

课表是一个三维结构，如果把课表看成一个表的话，那么它的主键为{课，星期，时间}。数据库的结构图如图 3-18 所示。

数据库操作的包包含五个类，如图 3-19 所示。

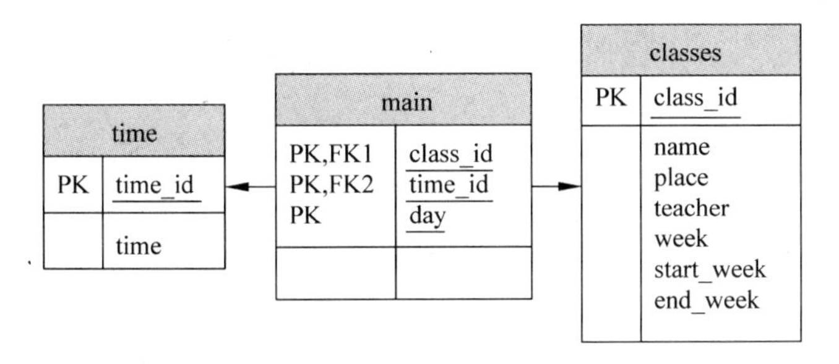

图 3-18　数据库结构图

com.now.timetable.db
ClassesDao.java
DBUtil.java
DBValue.java
MainDao.java
TimeDao.java

图 3-19　数据库操作类

DBUtil 用于创建表、初始化数据：

```
public class DBUtil extends SQLiteOpenHelper{

    public DBUtil(Context context) {
        //创建数据库
        super(context,DBValue.DB_NAME,null,1);
    }

    @Override
    public void onCreate(SQLiteDatabase db) {
        //建表
        db.execSQL(DBValue.Table_Time.CREATE_TABLE);
        db.execSQL(DBValue.Table_Classes.CREATE_TABLE);
        db.execSQL(DBValue.Table_Main.CREATE_TABLE);

        //插入初始数据
```

```
        initTimes(db);
    }

    private void initTimes(SQLiteDatabase db){
        insertTime("08:00",db);
        insertTime("10:00",db);
        insertTime("14:30",db);
        insertTime("16:30",db);
        insertTime("18:30",db);
        insertTime("20:30",db);
    }

    private void insertTime(String time,SQLiteDatabase db) {
        ContentValues cv = new ContentValues();
        cv.put(DBValue.Table_Time.TIME_COL,time);
        db.insert(DBValue.Table_Time.TABLE_NAME,null,cv);
    }


    @Override
    public void onUpgrade(SQLiteDatabase db,int oldVersion,int newVersion) {
    }

}
```

DBValue 用于封装每个表的字段、表名等静态数据。

为了节约篇幅，我省略了数据库操作方法的具体实现，这些代码在本书的代码中可以查看得到。

```
public class MainDao {

    private DBUtil dbUtil;
    private SQLiteDatabase db;
    private Context context;

    public MainDao(Context context) {
        dbUtil = new DBUtil(context);
        this.context = context;
    }

    /**
     * 把一节课插入 main 表中
     * 如果这节课不存在,插入到 classes 表中
     * 如果课名和地点一样,视为同一节课
     * @param vo
     */
    public void insertClass(ClassVo vo) {
    }
```

```
    /**
     * 更新 main 表(可能修改了时间和星期)和 classes 表
     * @param vo
     */
    public void updateClass(ClassVo vo){
    }

    /**
     * 只删除 main 表
     * @param classId
     */
    public void deleteClass(String classId) {
    }

    /**
     * 删除 main 表和 classes 表
     * @param classId
     */
    public void deleteClassBoth(String classId) {
    }

    /**
     * 查询一天课程
     * @param day
     * @return
     */
    public SortedTree selectClassByDay(int day) {
    }
}
```

ClassesDao 是包内可见,它并不给包外的类直接访问:

```
class ClassesDao {

    private DBUtil dbUtil;
    private SQLiteDatabase db;

    public ClassesDao(Context context) {
        dbUtil = new DBUtil(context);
    }

    /**
     * 增
     * @param vo
     * @return
     */
    public int insertClass(ClassVo vo) {
    }
```

```
    /**
     * 查,根据 class_id
     * @param vo
     * @return
     */
    public ClassVo selectClassById(String classId){

    }

    /**
     * 查,根据课名和地点
     * @param vo
     * @return
     */
    public int selectIdBynameAndPlace(String name,String place){
    }

    /**
     * 删
     * @param vo
     * @return
     */
    public void deleteClass(String classId) {
    }

    /**
     * 改
     * @param vo
     * @return
     */
    public void updateClass(ClassVo vo){
    }
}
```

TimeDao 为处理时间的数据库读取方法：

```
public class TimeDao {

    private DBUtil dbUtil;
    private SQLiteDatabase db;

    public TimeDao(Context context) {
        dbUtil = new DBUtil(context);
    }

    public void insertTime(String time) {
    }

    public void deleteTime(String time) {
    }
```

```
    public DbTimeVo selectAllTimes() {
    }

    public void changeTime(String oldTime,String newTime){
    }

    public String selectTimeById(int id) {
    }
}
```

实体类的包如图 3-20 所示。

```
com.now.timetable.entity
    ClassVo.java
    DbTimeVo.java
    MainVo.java
    NoteVo.java
    OptChild.java
    OptGroup.java
    Pref.java
    SortableItem.java
    SortedTree.java
    TimeVo.java
```

图 3-20　实体类

```
public class ClassVo implements Serializable,SortableItem{

    //classes 表
    private int classId;
    private int timeId;
    private String time = "";
    private String className = "";
    private String classPlace = "";
    private String teacher = "";
    private int week;
    private int startWeek;
    private int endWeek;
    private ArrayList<String> notes = new ArrayList<String>();

    //time 表
    private String hour = "??";
    private String min = "??";

    //main 表
    private int day;
```

界面设计

主界面 activity_timetable.xml 是一个 RelativeLayout 的布局，包含两个子元素，ListView 作为数据展示以及 LinearLayout 作为动作栏：

```
<RelativeLayout xmlns:android="http://schemas.android.com/apk/res/android"
    xmlns:tools="http://schemas.android.com/tools"
    android:layout_width="fill_parent"
    android:layout_height="fill_parent"
    android:background="@drawable/bcg" >

    <ListView
        android:id="@+id/activity_timetable_list"
        android:fadingEdge="none"
        android:clipToPadding="false"
        android:layout_marginLeft="10dip"
        android:divider="@color/transparent"
        android:dividerHeight="20dip"
        android:scrollbarThumbVertical="@color/transparent"
        android:layout_width="300dip"
        android:layout_height="fill_parent"
        android:paddingTop="42.0dip"
        />
    <LinearLayout
       android:layout_width="fill_parent"
       android:layout_height="48dip"
       android:background="@color/blur_black"
       android:gravity="center"
       >

       <Button
           android:id="@+id/activity_timetable_login_bt"
           style="@style/action_bar"
           android:background="@drawable/title_button_left_selector"
           android:text="@string/title_login"/>

       <TextView
           android:id="@+id/activity_timetable_week_bt"
           android:layout_width="140dip"
           android:layout_height="fill_parent"
           android:gravity="center"
           android:background="@color/transparent"
           android:text="@string/title_today"
           android:textColor="@color/white"
           android:textSize="16dip"/>

       <Button
           android:id="@+id/activity_timetable_opt_bt"
           style="@style/action_bar"
           android:background="@drawable/title_button_right_selector"
           android:text="@string/title_opt"/>

    </LinearLayout>

</RelativeLayout>
```

先来看 ListView 的设置：

```
android:divider = "@color/transparent"
android:dividerHeight = "20dip"
```

这两句话决定了列表的分隔符的风格，也就是透明的 20dp，效果是如图 3-21 所示。如果去掉这两行代码，效果如图 3-22 所示。

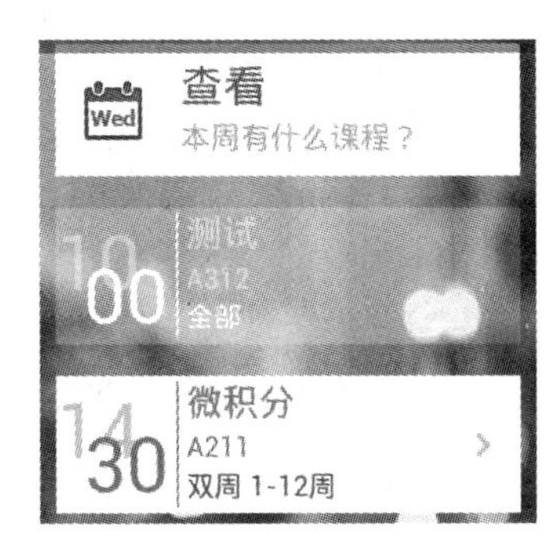

图 3-21　透明列表分隔符

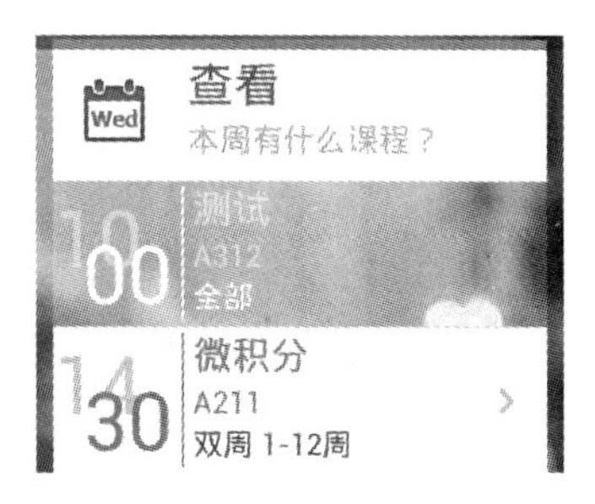

图 3-22　无定义分隔符

如果去掉这一句：

```
android:clipToPadding = "false"
```

ListView 的一部分会被顶端的控件所遮挡，如图 3-23 所示。

图 3-23　列表顶部被遮挡

因为设置了 paddingTop 的原因，整个列表会离顶部有 42dp 的距离，而代码

```
android:clipToPadding = "false"
```

的作用在于保持 ListView 在设置了 padding 的情况下依然可见(当然上面一层要是透明的)，如图 3-24 所示。

单击“查看”，会弹出一列菜单，如图 3-25 所示。

如果读者注意看的话会发现左上角和右上角的按钮有浅浅的竖线，形成凹陷的效果，如图 3-26 所示。

按下按钮时背景变白，如图 3-27 所示。

图 3-24　列表顶部没有被遮挡

图 3-25　星期选择菜单

图 3-26　凹陷效果

图 3-27　按下按钮的背景

这是因为在布局文件中分别设置了不同的 background。以左边登录按钮为例：

```
android:background = "@drawable/title_button_left_selector"
<?xml version = "1.0" encoding = "utf - 8"?>
< selector
  xmlns:android = "http://schemas.android.com/apk/res/android">
    < item android:state_pressed = "true"
android:drawable = "@drawable/title_button_left_pressed"/>
    < item android:drawable = "@drawable/title_button_left_normal"/>
</selector>
```

title_button_left_pressed 和 tittle_button_left_normal 是两张非常小的九格式图片。我用 Photoshop 把它们放大，如图 3-28 所示。

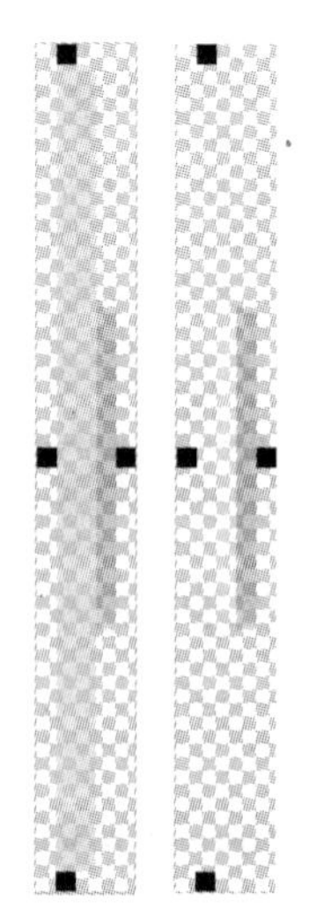

图 3-28　按钮背景位图

剩下的课程布局和大按钮布局也就没什么好说的了，就是把正确的文字、图片调到正确的大小，在 RelativeLayout 里放到正确的位置。

设置界面中，我使用了父列表有间隔线，子列表没有间隔线的 ExpandableListView，打开子列表有游标提示，如图 3-29 所示。

图 3-29　设置界面

这个效果的布局为(部分)：

```
android:listSelector = "@drawable/item_style_selector_transparent"
android:childDivider = "@color/transparent"
android:divider = "@drawable/list_divider_fade"
android:groupIndicator = "@color/transparent"
```

其中 listSelector 定义了列表被单击时的效果：

```
<?xml version = "1.0" encoding = "utf - 8"?>
< selector xmlns:android = "http://schemas.android.com/apk/res/android">
  < item android:state_pressed = "false" android:drawable = "@color/transparent"/>
  < item android:state_pressed = "true"  android:drawable = "@color/gray_line"/>
  < item android:drawable = "@color/transparent"/>
</selector >
```

divider 和 childDivider 分别定义的是父列表的分隔线和子列表的分隔线。

因为 ExpandableListView 默认会有一个指示的游标，如图 3-30 所示。

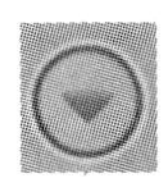

图 3-30　游标

所以我们用 groupIndicator 来指定自己的游标，也就是透明。如果去掉这一句效果如图 2-31 所示。

图 3-31　保留默认游标

修改上课信息看起来像个对话框，其实也是 Activity，如图 3-32 所示。

图 3-32　修改课程信息对话框

要做到这样的效果，可以在声明这个 activity 的同时声明它的 theme：

```
<activity
    android:name=".dialog.normal.EditClassDialog"
    android:theme="@style/blur_activity">
</activity>
```

这个 style 是这样写的：

```
<style name="blur_activity">
        <item name="android:windowNoTitle">true</item>
        <item name="android:windowAnimationStyle">
            @android:style/Animation.Dialog</item>
        <item name="android:windowBackground">@color/blur_black</item>
        <item name="android:windowIsTranslucent">true</item>
</style>
```

这个编辑框一节一节的效果是因为各自使用了不同的 background。最上方使用的 background 和最下方以及中间四个框都是不一样的，中间再用一条宽为 1dp 的 ImageView 隔开。

查看/编辑课表

应用主页面（查看课表）的逻辑比较简单，它负责显示并响应响应的事件（部分代码）：

```
private PopupWindow popupWindow;
private WeekPopAdapter popBaseAdapter;
/**
 * 保存了一周的课程，根据星期来获取这一天的课程
 */
private HashMap<Integer,SortedTree> classMap;
/**
 * 当前这一天的课程
 */
private SortedTree currentList;
private int day;
private static final String[] WEEK = {"Sun","Mon","Tue","Wed","Thu","Fri","Sar"};

private ListView listView;
private TimetableListAdapter adapter;

private TextView weekTv;

@Override
public void onCreate(Bundle savedInstanceState) {
    super.onCreate(savedInstanceState);
    setContentView(R.layout.activity_timetable);

    //载入课程数据
    initClasses();
    //获取当天课程
    currentList = getCurrentClass();

    //加载 ListView 数据
    listView = (ListView) findViewById(R.id.activity_timetable_list);
    adapter = new TimetableListAdapter(this,currentList,day);
```

```
    listView.setAdapter(adapter);

    /*
     * 添加 ListView 事件监听器,当按下"查看"按钮(第一行),弹出选择的 PopMenu;
     *         按下"增加"按钮(最后一行),弹出增加课程对话框
     */
    final Button optBt = (Button) findViewById(R.id.activity_timetable_opt_bt);
    listView.setOnItemClickListener(new OnItemClickListener(){
        @Override
        public void onItemClick(AdapterView<?> arg0,View arg1,int position,
                long arg3) {
            if(position == 0){
                //查看其他时间课程
                popupWindow.showAsDropDown(optBt, - 10,10);
            }else if(position == currentList.size() + 1){
                //增加课程
                startActivityForResult(new Intent(TimetableActivity.this,
                  EditClassDialog.class),Pref.ClassSet.REQUEST_CODE_ADD_CLASS);
            }
        }
    });

    //显示当前星期的标题栏
    weekTv = (TextView) findViewById(R.id.activity_timetable_week_bt);
    weekTv.setText(WEEK[day]);

    //初始化 PopMenu
     initPopupWindow();
     //开始通知服务
     startService();
}

//已在 layout 文件中定义 button 事件
public void startSettingActivity(View v){
    startActivity(new Intent(
         TimetableActivity.this,SettingActivity.class));
}
@Override
public void onActivityResult(int requestCode,int resultCode,Intent intent){
    //编辑/增加/删除课程回到该页面后,响应数据改变
    if(intent == null){
        return;
    }
    switch(requestCode){
    case Pref.ClassSet.REQUEST_CODE_SET_CLASS:
        if(resultCode == Pref.ClassSet.RESULT_CODE_CLASS_SET){
            int position = intent.getIntExtra(Pref.ClassSet.EXTRA_POSITION,0);
            ClassVo vo = (ClassVo) intent.getSerializableExtra(
                        Pref.ClassSet.EXTRA_CLASS);
            setClass(position,vo);
```

```
        }else if(resultCode == Pref.ClassSet.RESULT_CODE_CLASS_DELETE){
            int position = intent.getIntExtra(Pref.ClassSet.EXTRA_POSITION,0);
            deleteClass(position);
        }
        break;
    case Pref.ClassSet.REQUEST_CODE_ADD_CLASS:
        if(resultCode == Pref.ClassSet.RESULT_CODE_CLASS_ADD){
            ClassVo vo = (ClassVo) intent.getSerializableExtra(
                        Pref.ClassSet.EXTRA_CLASS);
            addClass(vo);
        }
        break;
    }
}
private void deleteClass(int position){
    //在数据库中删除
    ClassVo vo = (ClassVo) currentList.get(position);
    String id = "" + vo.getClassId();
    MainDao mainDao = new MainDao(this);
    mainDao.deleteClass(id);

    //更新 UI
    currentList.remove(position);
    adapter.notifyDataSetChanged();
}
private void addClass(ClassVo vo){
    vo.setDay(day);
    if(currentList.getItem() != null && currentList.contains(vo.getTime())){
        Toast.makeText(this,"该时间已经有课了",Toast.LENGTH_SHORT).show();
        return;
    }
    //插入数据库
    MainDao mainDao = new MainDao(this);
    mainDao.insertClass(vo);

    //更新 UI
    currentList.add(vo);
    adapter.notifyDataSetChanged();
}
private void setClass(int position,ClassVo vo){
    //修改数据库
    MainDao mainDao = new MainDao(this);
    mainDao.updateClass(vo);

    //更新 UI
    currentList.set(position,vo);
    adapter.notifyDataSetChanged();
}
```

编辑课表的 Activity 逻辑较为简单：如果 onCreate 时有数据（修改课程），就把数据读出来填入编辑框中；就是当单击“确定”按钮时，把数据打包起来并回传回去。

此外，编辑课程框中有两处动画效果：

```
//勾选"选择"时,选择周控件滑入
public void slideInWeekSelector(){
    TranslateAnimation anima = new TranslateAnimation(300,0,0,0);
    anima.setFillAfter(true);
    anima.setDuration(500);
    weekSelector.startAnimation(anima);
}

//取消勾选"选择"时,选择周控件滑出
public void slideOutWeekSelector(){
    TranslateAnimation anima = new TranslateAnimation(0,300,0,0);
    anima.setFillAfter(true);
    anima.setDuration(500);
    weekSelector.startAnimation(anima);
}

//初始化时间选择控件时,利用线程播放翻滚的动画
private void initTimeWheel(){
    timeWheel = (WheelView) findViewById(R.id.dialog_edit_class_time_wheel);
    timeWheel.setAdapter(new StringWheelAdapter(timeList));
    timeWheel.setCyclic(true);
    timeWheel.TEXT_SIZE = 60;
    timeWheel.setVisibleItems(1);
    timeWheel.setCurrentItem(timeIndex);
    new Thread(){
        public void run(){
            try {
                sleep(100);
            } catch (InterruptedException e) {
                //TODO Auto - generated catch block
                e.printStackTrace();
            }
            //播放翻滚动画,刚好又回到正确的位置
            timeWheel.scroll(timeList.size(),700);
        }
    }.start();
}
```

配置页面

配置界面中，如图 3-33 所示，子列表的叉叉看起来像图标，但其实就是一个 TextView 而已，如图 3-34 所示（谁说 TextView 不能有单击事件的?）。它的子元素的布局就是三个简单的 TextView。

配置界面是一个可开闭的 ExpandableListView，这个控件我们将在下一章详细讲解，这里不多加赘述。

图 3-33 TextView 作为按钮

图 3-34 配置界面

获取子视图的代码为：

```
@Override
public View getChildView(final int groupPosition,final int childPosition,boolean isLastChild,
View convertView,ViewGroup parent) {
    ChildHolder holder;
    if(convertView == null){
        convertView = LayoutInflater.from(mContext).inflate(
                R.layout.list_setting_child,null);
        holder = new ChildHolder();
        holder.hint = (TextView) convertView.findViewById(
                R.id.list_setting_child_hint_tv);
        holder.value = (TextView) convertView.findViewById(
                R.id.list_setting_child_value_tv);
        holder.addition = (TextView) convertView.findViewById(
                R.id.list_setting_child_addition_tv);
        convertView.setTag(holder);
    }else{
        holder = (ChildHolder)convertView.getTag();
    }
    //设置内容
    OptChild child = (OptChild) mGroups.get(groupPosition).
            getChildren().get(childPosition);
    holder.hint.setText(child.getHint());
    String value = child.getValue();
    if(groupPosition == 2){
        if(value.equals("0")){
            holder.value.setText("不提醒");
        }else{
            holder.value.setText(value + "分钟");
```

```
        }
    }else{
        holder.value.setText(value);
    }
    String addition = child.getAddition();
    if(addition.equals("") || addition == null){
        holder.addition.setVisibility(View.INVISIBLE);
    }else{
        holder.addition.setText(addition);
    }
    //添加事件
    if(groupPosition == 0){
        holder.addition.setOnClickListener(new OnClickListener(){
            @Override
            public void onClick(View v) {
                mGroups.get(groupPosition).
                        removeChildInOrder(childPosition);
                notifyDataSetChanged();
            }
        });
    }
    return convertView;
}
```

桌面小插件

桌面小插件沿用了课程列表的布局，有三种配色可供选择，如图 3-35 所示。

图 3-35　桌面小插件

更新视图的逻辑为：如果下一节课的上课时间晚于现在时间，那么更新插件信息；如果当前时间晚于任何一节课的时间，则不显示上课内容。

```
public void updateView(Context context,AppWidgetManager wManager)
{
    ClassVo nextClass = packInfo(context);
    int layout = new UserPref(context).getAppearance();
    switch(layout){
    case UserPref.FLAG_BLUR:
        views = new RemoteViews(
            context.getPackageName(),R.layout.layout_widget_blur);
        break;
    case UserPref.FLAG_BLUE:
```

```
            views = new RemoteViews(
                context.getPackageName(),R.layout.layout_widget_blue);
            break;
        case UserPref.FLAG_WHITE:
        default:
            views = new RemoteViews(
                context.getPackageName(),R.layout.layout_widget_white);
            break;
        }
        if(nextClass != null){
            views.setTextViewText(R.id.layout_widget_place_tv,
                nextClass.getClassPlace());
            views.setTextViewText(R.id.layout_widget_class_tv,
                nextClass.getClassName());
            TimeVo time = new TimeVo(nextClass.getTime(),true);
            views.setTextViewText(R.id.layout_widget_hour_tv,
                time.getHourString());
            views.setTextViewText(R.id.layout_widget_min_tv,time.getMinString());
            views.setTextViewText(R.id.layout_widget_week_tv,
                nextClass.getWeekDisplay());
            wManager.updateAppWidget(new ComponentName(
                context,MyAppWidgetProvider.class),views);
        }else{
            views.setTextViewText(R.id.layout_widget_place_tv,"");
            views.setTextViewText(R.id.layout_widget_class_tv,"今天没课了");
            TimeVo time = new TimeVo();
            views.setTextViewText(R.id.layout_widget_hour_tv,
                time.getHourString());
            views.setTextViewText(R.id.layout_widget_min_tv,time.getMinString());
            views.setTextViewText(R.id.layout_widget_week_tv,"");
            wManager.updateAppWidget(new ComponentName(
                 context,MyAppWidgetProvider.class),views);
        }
    }
    private ClassVo packInfo(Context context)
    {
        MainDao mainDao = new MainDao(context);
        SortedTree classList = mainDao.selectClassByDay(getWeekOfDate());
        TimeUtil timeManage = new TimeUtil(classList);
        String startWeek = new UserPref(context).getDate();
        ClassVo vo = timeManage.getNextClass(20);
        if(vo == null){
            return null;
        }
        boolean isClass = ifClass(vo.getWeek(),vo.getStartWeek(),vo.getEndWeek(),
            timeManage.getCurrentWeek(startWeek));
        if(isClass){
            return vo;
        }else{
            return null;
        }
    }
```

```
private boolean ifClass(int flag, int startWeek, int endWeek, int currentWeek)
{
    switch(flag)
    {
    case DBValue.Table_Classes.ROW_ALL:
        break;
    case DBValue.Table_Classes.ROW_SINGLE:
        if(currentWeek % 2 != 1)
            return false;
        break;
    case DBValue.Table_Classes.ROW_DOUBLE:
        if(currentWeek % 2 != 0)
            return false;
        break;
    }
    if(startWeek + endWeek == 0){
        return true;
    }
    if(currentWeek >= startWeek && currentWeek <= endWeek){
        return true;
    }
    return false;
}
/**
 * Sunday 0 Saturday 6
 * @return
 */
private int getWeekOfDate()
{
    Date dt = new Date();
    Calendar cal = Calendar.getInstance();
    cal.setTime(dt);

    int w = cal.get(Calendar.DAY_OF_WEEK) - 1;
    if (w < 0)
        w = 0;

    return w;
}
```

定时通知的实现

定时通知的实现为通过一个 Service 设定定时器，每 1 分钟查询一次上课内容的变化。

```
private void setNoti(int iconId, ClassVo vo)
{
        Intent notifyIntent = new Intent(this, TipDailog.class);
        notifyIntent.setFlags(Intent.FLAG_ACTIVITY_NEW_TASK);
        notifyIntent.putExtra(Pref.Notification.EXTRA_CLASS, vo.getClassName());
```

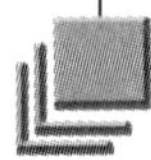

```
        notifyIntent.putExtra(Pref.Notification.EXTRA_PLACE,vo.getClassPlace());
        notifyIntent.putExtra(Pref.Notification.EXTRA_TIME,vo.getTime());
        PendingIntent appIntent = PendingIntent.getActivity(
             NotificationService.this,0,notifyIntent,0);

        Notification myNoti = new Notification();
        myNoti.icon = iconId;
        myNoti.tickerText = vo.getClassName();
        myNoti.defaults = Notification.DEFAULT_SOUND;
        myNoti.flags = Notification.FLAG_AUTO_CANCEL;
        myNoti.setLatestEventInfo(NotificationService.this,"下节课即将开始",
             vo.getClassName(),appIntent);
        myNotiManager.notify(0,myNoti);
    }
    private ClassVo packInfo()
    {
        MainDao mainDao = new MainDao(this);
        SortedTree classList = mainDao.selectClassByDay(getWeekOfDate());
        TimeUtil timeManage = new TimeUtil(classList);
        UserPref pref = new UserPref(this);
        String startWeek = pref.getDate();
        //如果当前时间后的n分钟(用户设定)内等于上课时间,则弹出通知
        ClassesVo vo = timeManage.getNextClassExactly(pref.getNoti());
        if(vo == null){
            return null;
        }
        boolean isClass = ifClass(vo.getWeek(),vo.getStartWeek(),
                vo.getEndWeek(),timeManage.getCurrentWeek(startWeek));
        if(isClass){
            return vo;
        }else{
            return null;
        }
    }
private boolean ifClass(int flag,int startWeek,int endWeek,int currentWeek)
{
    switch(flag)
    {
    case DBValue.Table_Classes.ROW_ALL:
        break;
    case DBValue.Table_Classes.ROW_SINGLE:
        if(currentWeek % 2 != 1)
            return false;
        break;
    case DBValue.Table_Classes.ROW_DOUBLE:
        if(currentWeek % 2 != 0)
            return false;
        break;
    }
    if(currentWeek >= startWeek && currentWeek <= endWeek){
```

```
            return true;
        }
        return false;
    }
    private int getWeekOfDate()
    {
        Date dt = new Date();
        Calendar cal = Calendar.getInstance();
        cal.setTime(dt);

        int w = cal.get(Calendar.DAY_OF_WEEK) - 1;
        if (w < 0)
          w = 0;

        return w;
    }
```

第四章　TODO

用滑动手势打造用户体验

清单

Demo 代码：

\demo\Demo_todo

实例代码：

\source_codes\to - do

Target SDK：

Android 1.6

第一节　产 品 介 绍

待办事项这一类的应用应该是很多很多了。既然已经这么多了，那么无所谓再多一个(囧)。事实上，这一类应用最重要的功能之一就是把已经过期的事件重新安排回去。因为我们总是会被其他的事情打断，以至于事情没有在预想的时间内完成。本应用主要学习自制滑动列表以及 Animation 的使用。滑动列表的逻辑会比较复杂，也是本章的重点。

这一类产品做得好的有 Any. DO，也是我的模仿对象，如图 4-1 所示。

图 4-1　Any. DO 应用

需求分析

日程应用有以下几个功能：

- 查看一周日程；
- 编辑日程；
- 过期事件列表。

界面设计

界面风格模仿 Any. DO 的清爽与简洁，使用蓝白色的色调，如图 4-2 所示。

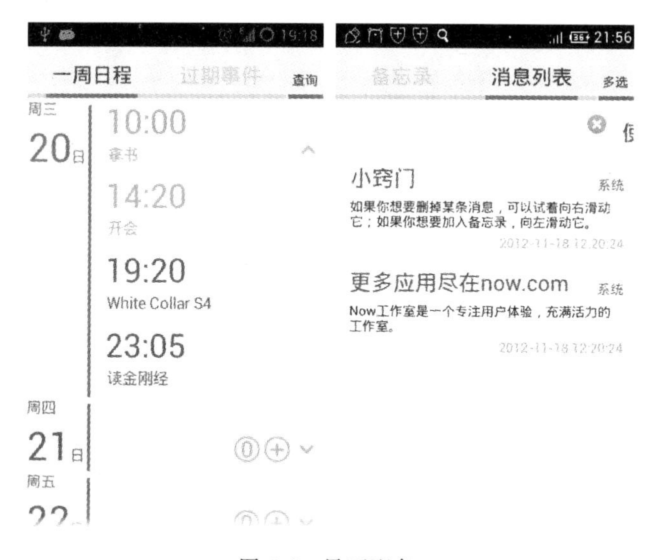

图 4-2　界面配色

用户体验设计

在这个应用里面，我使用了可以滑动子元素的 ListView。同时借鉴了三星通讯录滑动拨号和 Any. DO 的滑动删除功能，如图 4-3 所示。

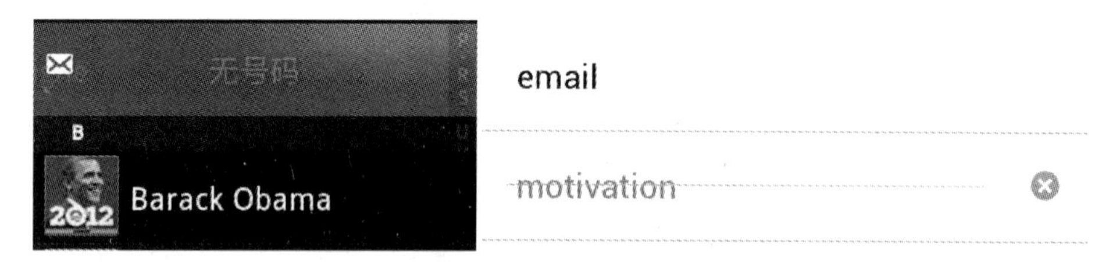

图 4-3　三星通讯录(左)和 Any. DO(右)

向左滑动时，导航栏上的"一周日程"会变蓝，提示用户这个操作将会把该日程重新安排至本周；松开手时，"一周日程"会变成深蓝，提示用户该日程已经加入到本周，如图 4-4 所示。

向右滑动时，并不会直接删除，而是露出剩余的一小截并显示删除的按钮，用户按下删除按钮才会把日程删除，如图 4-5 所示。这就相当于提醒用户"确定要删除该日程?"。

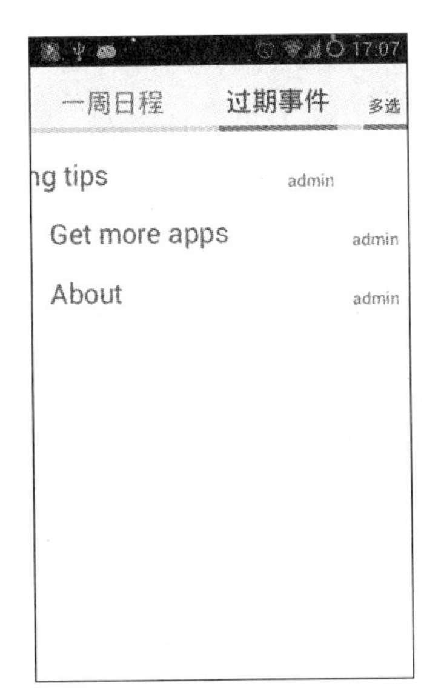

图 4-4　向左滑动列表

图 4-5　删除事项

当然，这些用户体验都是建立在正常的操作之上的，单击右上角"多选"按钮则会显示选择框，如图 4-6 所示。

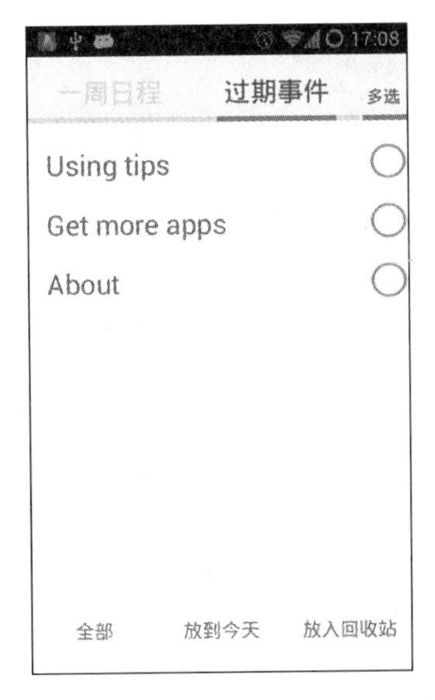

图 4-6　选择事项

第二节　手风琴 ExpandableListView

配置布局文件

我们在前面一章使用过 ExpandableListView，跟 ListView 一样，使用适配器来配置数据。先准备三个布局文件，先是定义一个 ExpandableListView 的 activity 布局文件 activity_list. xml：

```
<RelativeLayout xmlns:android="http://schemas.android.com/apk/res/android"
    xmlns:tools="http://schemas.android.com/tools"
    android:layout_width="fill_parent"
    android:layout_height="fill_parent">

    <ExpandableListView
        android:id="@+id/expandableLV"
        android:layout_width="fill_parent"
        android:layout_height="wrap_content"
        android:groupIndicator="@android:color/transparent"
        android:divider="@android:color/transparent"
        android:childDivider="@android:color/transparent"
        android:layout_alignParentLeft="true"
        android:layout_alignParentTop="true">
    </ExpandableListView>

</RelativeLayout>
```

再定义父容器(也就是未展开之前的ListView)的layout(layout_group.xml):

```
<?xml version="1.0" encoding="utf-8"?>
<LinearLayout xmlns:android="http://schemas.android.com/apk/res/android"
    android:layout_width="fill_parent"
    android:layout_height="wrap_content"
    android:orientation="vertical">
    <LinearLayout
      android:layout_width="fill_parent"
        android:layout_height="wrap_content"
        android:orientation="horizontal"
        android:layout_marginTop="5dp"
        android:layout_marginBottom="5dp"
        android:gravity="center_vertical">

        <TextView
            android:id="@+id/name"
            android:layout_width="wrap_content"
            android:layout_height="wrap_content"
            android:textSize="16dp"
            android:drawableLeft="@drawable/ic_indicator"
            android:drawablePadding="5dp"/>

    </LinearLayout>
</LinearLayout>
```

看起这个布局似乎多了一层没用的LinearLayout(最外面一层),实则不然。因为ListView的子元素(也就是上面的根元素)的layout_height会强制变为wrap_content。

子容器layout(layout_child.xml):

```
<?xml version="1.0" encoding="utf-8"?>
<LinearLayout xmlns:android="http://schemas.android.com/apk/res/android"
    android:layout_width="fill_parent"
    android:layout_height="wrap_content"
    android:orientation="vertical" >
    <LinearLayout
        android:layout_width="fill_parent"
        android:layout_height="60dp"
        android:gravity="center_vertical"
        android:orientation="horizontal" >

        <ImageView
            android:id="@+id/head"
            android:layout_width="50dp"
            android:layout_height="50dp"/>

        <LinearLayout
            android:layout_width="wrap_content"
            android:layout_height="fill_parent"
            android:gravity="center_vertical"
            android:layout_marginLeft="5dp"
            android:orientation="vertical" >

            <TextView
                android:id="@+id/name"
                android:layout_width="wrap_content"
                android:layout_height="wrap_content"
                android:textSize="16dp"/>

            <TextView
                android:id="@+id/word"
                android:layout_width="wrap_content"
                android:layout_height="wrap_content"
                android:textSize="16dp"
                android:textColor="#dd666666"/>

        </LinearLayout>

    </LinearLayout>
</LinearLayout>
```

使用适配器

由于是二级列表的关系，所以适配器的数据是 ArrayList 嵌 ArrayList 的结构：

```
public class GroupVo {

    private String name;
    private ArrayList<ChildVo> children;
```

```
    }
    public class ChildVo {

        private int head;
        private String name;
        private String jabber;
    }
    /*
     * 获取数据
     */
    private ArrayList<GroupVo> getGroup()
```

ExpandableListView 的适配器本质上跟 ListView 是一样的，只不过 ExpandableListView 的适配器区分了 child 跟 group：

```
public class ExpandableAdapter extends BaseExpandableListAdapter{

    private ArrayList<GroupVo> group;
    private Context context;

    public ExpandableAdapter(ArrayList<GroupVo> group,Context context){
        this.group = group;
        this.context = context;
    }

    @Override
    public int getChildrenCount(int groupPosition) {
        return group.get(groupPosition).getChildren().size();
    }

    @Override
    public View getChildView(int groupPosition,int childPosition,
            boolean isLastChild,View convertView,ViewGroup parent) {
        return convertView;
    }

    @Override
    public int getGroupCount() {
        return group.size();
    }

    @Override
    public View getGroupView(int groupPosition,boolean isExpanded,
            View convertView,ViewGroup parent) {
        return convertView;
    }

    @Override
    public boolean isChildSelectable(int groupPosition,int childPosition) {
```

```
            //这里要 return true,ListView 的子列才可以选择
            return true;
        }

        @Override
        public Object getGroup(int groupPosition) {
            return null;
        }

        @Override
        public Object getChild(int groupPosition,int childPosition) {
            return null;
        }

        @Override
        public long getGroupId(int groupPosition) {
            return 0;
        }

        @Override
        public long getChildId(int groupPosition,int childPosition) {
            return 0;
        }

        @Override
        public boolean hasStableIds() {
            //TODO Auto - generated method stub
            return false;
        }
    }
```

第三节　动画 Animation

Animation 可以实现多种效果的动画，我们在上一节中使用的 android. R. anim 就是系统自己定义的一个 Animation。

Animation 有以下两种模式。

- Tween Animation　通过对场景里的对象不断做图像变换（平移、缩放、旋转）产生动画效果。
- Frame Animation　顺序播放事先做好的图像。

幻灯片 TweenAnimation

在 TweenAnimation 中，我们可以设置四种动画效果。

- alpha　渐变透明度。
- scale　渐变尺寸伸缩。

- translate　画面转换位置移动。
- rotate　画面转移旋转。

TweenAnimation 共同的节点属性如表 4-1 所示。

表 4-1　TweenAnimation 共同节点

属　性	类型	解　释
android:duration	long	动画持续时间
android:fillAfter	boolean	为 true 时，动画结束后会应用该动画转化
android:fillBefore	boolean	为 true 时，动画结束前会应用该动画转化
android:interpolator		指定一个动画插入器
android:repeatCount	int	动画的重复次数
android:repeatMode	int	重复模式：1 为重新开始；2 为逆效果
android:startOffset	long	动画间的时间间隔
android:zAdjustment	int	动画的层：0 保持不变；1 保持最上层；2 保持最下层

首先，我们来看伸缩效果的 Animation，如表 4-2 所示。

表 4-2　scaleAnimation 节点

属　性	类　型	解　释
android:fromXScale android:fromYScale	float float	动画开始时，X、Y 坐标上的伸缩尺度； 小于 1 表示缩小，大于 1 表示放大
android:toXScale android:toYScale	float float	动画结束时，X、Y 坐标上的伸缩尺寸； 小于 1 表示缩小，大于 1 表示放大
android:pivotX android:pivotX	String String	相对于物件的 X、Y 坐标的开始位置，取值 0～100%； 50%为物件的 X 或 Y 方向坐标上的中点位置

定义 TweenAnimation 的 xml 文件需要放在 res/anim 文件夹下，如图 4-7 所示。

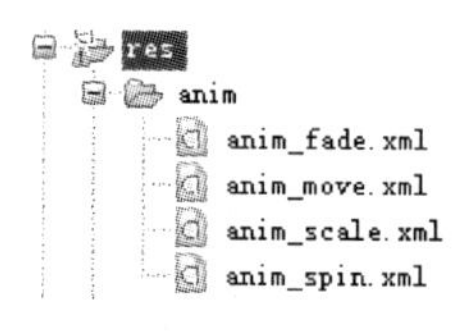

图 4-7　动画文件的存放

缩放效果 Animation 示例(anim_iv.xml)：

```
<?xml version = "1.0" encoding = "utf - 8"?>
< set android:shareInterpolator = "false"
    xmlns:android = "http://schemas.android.com/apk/res/android">
    < scale
        android:fromXScale = "0.5"
        android:toXScale = "1.5"
        android:fromYScale = "1.5"
        android:toYScale = "0.5"
        android:pivotX = "50 % "
```

```
            android:pivotY = "50 % "
            android:fillAfter = "true"
            android:duration = "10000"/>
</set>
```

在 Java 代码中播放动画：

```
public void scale(View v){
        ImageView iv = (ImageView) findViewById(R.id.anim_iv);
      iv.setImageResource(R.drawable.ic_launcher);
        Animation anim = AnimationUtils.loadAnimation(this,R.anim.anim_scale);
        iv.startAnimation(anim);
}
```

效果如图 4-8 所示。

图 4-8　伸缩 Animation 效果示意

此外，如果不使用 xml 文件，也可以直接在 Java 代码中定义，效果是一样的：

```
ImageView iv = (ImageView) findViewById(R.id.anim_iv);
//fromX,toX,fromY,toY,pivotX,pivotY
Animation anim = new ScaleAnimation(0.5f,1.5f,1.5f,0.5f,0.5f,0.5f);
anim.setDuration(10000);
anim.setFillAfter(true);
iv.startAnimation(anim);
```

接下来是透明变化的 Animation，它的节点属性如表 4-3 所示。

表 4-3　透明变化 Animation 节点

属　　性	类　　型	解　　释
android:fromAlpha	long	开始的透明度，0 为全透明
android:toAlpha	long	结束的透明度，1 为完全不透明

我们来试一下透明变化的效果：

```
<?xml version = "1.0" encoding = "utf - 8"?>
< set android:shareInterpolator = "false"
     xmlns:android = "http://schemas.android.com/apk/res/android">
     < alpha
```

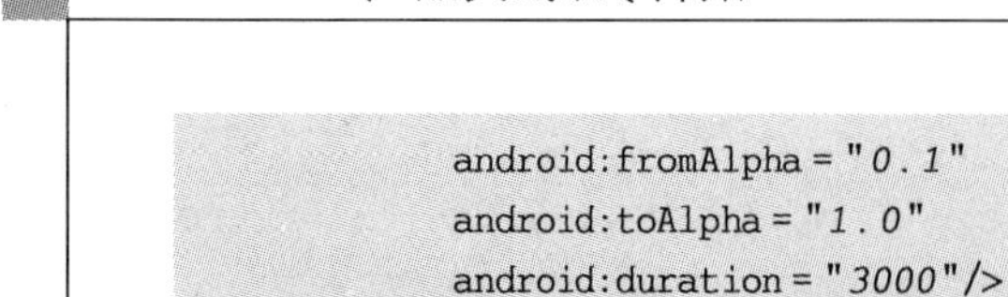

```
        android:fromAlpha = "0.1"
        android:toAlpha = "1.0"
        android:duration = "3000"/>
</set>
```

在 Activity 中设置控件的动画效果：

```
ImageView iv = (ImageView) findViewById(R.id.anim_iv);
Animation anim = AnimationUtils.loadAnimation(this,R.anim.anim_fade);
iv.startAnimation(anim);
```

效果如图 4-9 所示。

图 4-9　透明 Animation 效果示意

平移的 Animation 属性如表 4-4 所示。

表 4-4　平移 Animation 节点属性

属　　性	类　　型	解　　释
android:fromXDelta android:fromYDelta	int int	动画开始时的坐标
android:toXDelta android:toYDelta	int int	动画结束时的坐标

平移效果 Animation 示例：

```
<?xml version = "1.0" encoding = "utf - 8"?>
< set android:shareInterpolator = "false"
    xmlns:android = "http://schemas.android.com/apk/res/android">
    < translate
        android:fromXDelta = "0"
        android:fromYDelta = "0"
        android:toXDelta = "200"
        android:toYDelta = "30"
        android:duration = "10000"/>
</set>
```

旋转的 Animation 属性如表 4-5 所示。

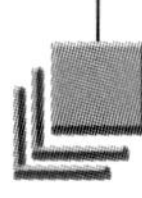

表 4-5　旋转 Animation 节点属性

属　性	类　型	解　释	备　注
android:fromDegrees	int	动画开始时控件的旋转角度	(from 负数——to 正数:顺时针旋转) (from 负数——to 负数:逆时针旋转) (from 正数——to 正数:顺时针旋转) (from 正数——to 负数:逆时针旋转)
android:toDegrees	int	动画结束时控件的旋转角度	
android:pivotX	String	动画相对于控件的坐标开始位置,取值0～100%	50%为控件的 x 或 y 坐标上的总代理位置
android:pivotY	String		

旋转效果 Animation 示例:

```
<?xml version = "1.0" encoding = "utf - 8"?>
< set android:shareInterpolator = "false"
    xmlns:android = "http://schemas.android.com/apk/res/android">
    < rotate
        android:fromDegrees = "0"
        android:toDegrees = " - 45"
        android:pivotX = "50 % "
        android:pivotY = "50 % "
        android:duration = "10000"/>
</set >
```

显示的效果如图 4-10 所示。

图 4-10　旋转 Animation 效果示意

电影胶片 FrameAnimation

如果是在 Web 中,要设置一张 gif 动态图片非常简单,就跟普通的图片无异。但是在 Android 里面就不行了(Android 没有提供官方控件,但是有第三方控件支持),我们需要使用 Animation 来一张一张地播放动态图片。也就是说我们需要把 gif 动态图片截成一张张,然后再一张一张地切换。

anim_frame.xml 的代码如下:

```
<?xml version = "1.0" encoding = "UTF - 8"?>
< animation - list android:oneshot = "false"
```

```
    xmlns:android="http://schemas.android.com/apk/res/android">
    <item android:duration="150" android:drawable="@drawable/frame_01"/>
    <item android:duration="150" android:drawable="@drawable/frame_02"/>
    <item android:duration="150" android:drawable="@drawable/frame_03"/>
    <item android:duration="150" android:drawable="@drawable/frame_04"/>
    <item android:duration="150" android:drawable="@drawable/frame_05"/>
    <item android:duration="150" android:drawable="@drawable/frame_06"/>
    <item android:duration="150" android:drawable="@drawable/frame_07"/>
</animation-list>
```

注意：这个 xml 文件不能放在 anim 文件夹下，而是要放到 drawable 文件夹中。并且开始这个 animation 的方法也不一样：

```
public void frame(View v){
    ImageView iv = (ImageView) findViewById(R.id.anim_iv);
    iv.setImageResource(R.drawable.anim_frame);
    AnimationDrawable animationDrawable = (AnimationDrawable) iv.getDrawable();
     animationDrawable.start();
}
```

第四节　功能实现

界面实现

这个应用的界面风格为简洁的蓝白配色，主界面如图 4-11 所示。

图 4-11　界面效果

它的布局框架为一个线性布局，里面包含一个相对布局（导航栏）和一个卡片布局（列表）：

```
<LinearLayout
    android:layout_width = "fill_parent"
    android:layout_height = "fill_parent"
    android:orientation = "vertical"

    xmlns:android = "http://schemas.android.com/apk/res/android">
     <RelativeLayout
          android:id = "@+ id/relativeLayout1"
          android:layout_height = "55dip"
          android:layout_width = "fill_parent"
          android:background = "@color/gray_tiny" >
     </RelativeLayout>
     <FrameLayout
         android:id = "@+ id/activity_list_root"
         android:layout_width = "fill_parent"
         android:layout_height = "fill_parent"
         android:layout_below = "@+ id/relativeLayout1"
         android:layout_alignParentLeft = "true" >
     </FrameLayout>
</LinearLayout>
```

编辑日程也是同样的风格，如图 4-12 所示。

图 4-12　编辑日程

导航栏之下是一个传统的垂直 LinearLayout(框架代码)：

```
<LinearLayout xmlns:android = "http://schemas.android.com/apk/res/android"
    android:layout_width = "fill_parent"
    android:layout_height = "fill_parent"
    android:gravity = "center"
    android:orientation = "vertical" >
```

```
<!--导航栏-->
<RelativeLayout
    android:id="@+id/relativeLayout1"
    android:layout_height="55dip"
    android:layout_width="fill_parent"
    android:background="@color/gray_tiny">
</RelativeLayout>
<!--时间选择-->
<LinearLayout
    android:layout_width="fill_parent"
    android:layout_height="wrap_content"
        android:gravity="center"
    android:orientation="horizontal">
</LinearLayout>

<!--分隔线-->
<ImageView style="@style/divider"/>

<!--标题输入-->
<EditText
    android:id="@+id/activity_edit_item_title_et"
    style="@style/input"
    android:layout_height="40dip"
    android:hint="@string/hint_title">
</EditText>

<!--分隔线-->
<ImageView style="@style/divider"/>

<!--内容输入-->
<EditText
    android:id="@+id/activity_edit_item_content_et"
    style="@style/input"
    android:layout_height="120dip"
    android:hint="@string/hint_content">
</EditText>

<!--分隔线-->
<ImageView style="@style/divider"/>

<!--日期选择-->
<LinearLayout
    android:layout_width="320dip"
    android:layout_height="60dip"
    android:gravity="center">
</LinearLayout>

<!--按钮-->
<LinearLayout
        android:layout_width="fill_parent"
```

```
            android:layout_height = "fill_parent"
            android:gravity = "bottom" >
    </LinearLayout >

</LinearLayout >
```

关于里面两个列表以及导航栏的具体实现,我们会接下来讲解。

导航栏的实现

主界面使用了一个自定义的导航栏,如图 4-13 所示。

图 4-13 导航栏

这一部分的布局如下:

```
< RelativeLayout
    android:id = "@+ id/relativeLayout1"
    android:layout_height = "55dip"
    android:layout_width = "fill_parent"
    android:background = "@color/gray_tiny" >
    <! -- 灰色的线 -- >
    < ImageView
        android:id = "@+ id/navi_line_iv"
        android:layout_height = "5dip"
        android:layout_width = "280dip"
        android:layout_alignParentBottom = "true"
        android:src = "@color/gray_line"
    />
    <! -- 蓝色导航线 -- >
    < ImageView
        android:id = "@+ id/navi_moving_line_iv"
        android:layout_height = "5dip"
        android:layout_width = "100dip"
        android:layout_alignParentBottom = "true"
        android:layout_alignParentLeft = "true"
        android:layout_marginLeft = "20dip"
        android:src = "@color/blue_light"
        />
    <! -- 多选按钮下的蓝线 -- >
    < ImageView
        android:id = "@+ id/navi_muti_line_iv"
        android:layout_height = "5dip"
        android:layout_width = "38dip"
        android:layout_alignParentBottom = "true"
```

```
        android:layout_alignParentRight = "true"
        android:src = "@color/blue_light"
        />
    <!-- -周日程按钮 -->
    <Button
        android:id = "@+id/title_memo_bt"
        style = "@style/navi_bar_text"
        android:textColor = "@color/blue"
        android:layout_above = "@+id/navi_line_iv"
        android:layout_alignParentLeft = "true"
         android:onClick = "clickWeekButton"
        android:text = "@string/title_memo"/>
    <!-- 过期事件按钮 -->
    <Button
        android:id = "@+id/title_msg_bt"
        style = "@style/navi_bar_text"
        android:layout_above = "@+id/navi_line_iv"
        android:layout_toRightOf = "@+id/title_memo_bt"
        android:onClick = "clickMsgButton"
        android:text = "@string/title_msg_list"/>
    <!-- 多选按钮 -->
    <Button
        android:id = "@+id/activity_list_muti_bt"
        style = "@style/navi_muti_bt"
        android:layout_above = "@+id/navi_line_iv"
        android:text = "@string/title_search"
     />

</RelativeLayout>
```

导航栏之下就是列表主体了。因为我们需要在两个不同列表中进行切换，所以使用了一个 FrameLayout：

```
<FrameLayout
    android:id = "@+id/activity_list_root"
    android:layout_width = "fill_parent"
    android:layout_height = "fill_parent"
    android:layout_above = "@+id/activity_list_move_lt"
    android:layout_below = "@+id/relativeLayout1"
    android:layout_alignParentLeft = "true" >

    <!-- 过期事件列表(包含隐藏的三个按钮) -->
    <RelativeLayout
        android:id = "@+id/activity_list_msg_lt"
        android:visibility = "gone"
        android:layout_height = "fill_parent"
        android:layout_width = "fill_parent">

        <edu.csu.icp.view.FlingExpandListView
```

```
            android:id = "@ + id/activity_list_msg_list"
            android:layout_width = "fill_parent"
            android:layout_height = "fill_parent"
            android:background = "@drawable/list_selector"
            android:listSelector = "@drawable/list_selector"
            android:scrollbarThumbVertical = "@drawable/ic_scroller"
            android:divider = "@color/transparent"
            android:drawSelectorOnTop = "false"
            android:groupIndicator = "@color/transparent"/>
            <! -- 按下多选按钮后,弹出三个选项按钮 -->
        < LinearLayout
            android:id = "@ + id/activity_list_move_lt"
            android:layout_height = "45dip"
            android:layout_width = "fill_parent"
            android:visibility = "gone"
            android:layout_alignParentBottom = "true" >

            < Button
                android:id = "@ + id/activity_list_select_all_bt"
                style = "@style/muti_select_text"
                android:text = "@string/select_all"/>

            < Button
                android:id = "@ + id/activity_list_move_bt"
                style = "@style/muti_select_text"
                android:text = "@string/move_msg"/>

            < Button
                android:id = "@ + id/activity_list_remove_bt"
                style = "@style/muti_select_text"
                android:text = "@string/remove_msg"/>
        </LinearLayout >
    </RelativeLayout >

    < ExpandableListView
        android:id = "@ + id/activity_list_week_list"
        android:layout_width = "fill_parent"
        android:layout_height = "fill_parent"
        android:background = "@drawable/list_selector"
        android:listSelector = "@drawable/list_selector"
        android:scrollbarThumbVertical = "@drawable/ic_scroller"
        android:divider = "@color/transparent"
        android:drawSelectorOnTop = "false"
        android:groupIndicator = "@color/transparent"/>
</FrameLayout >
```

切换周视图和过期日程：

```
public void clickMsgButton(View v){
    /*
     * 如果现在的视图不是过期日程
     */
    if(tabFlag != FLAG_MSG){
        //还原状态
        msgListView.getMsgListView().onItemBack();
        msgListView.getMsgAdapter().onMutiDisappear();
        //改变颜色
        weekBt.setTextColor(color_grayLine);
        msgBt.setTextColor(color_blue);

        //开始移动蓝色导航线的 Animation
        movingLineIv.startAnimation(mSlidingLeftAnim);
        new Thread(){
            public void run(){
                try {
                    sleep(SLIDE_DURATION);
                } catch (InterruptedException e) {
                    e.printStackTrace();
                }
                /*
                 * 设置蓝色导航线的新位置
                 * 将状态改变为过期日程
                 */
                handler.sendEmptyMessage(FLAG_MSG);
                tabFlag = FLAG_MSG;
            }
        }.start();
    }
}

public void clickWeekButton(View v){
    if(tabFlag != FLAG_WEEK){
        //还原状态
        msgListView.getMsgListView().onItemBack();
        msgListView.getMsgAdapter().onMutiDisappear();
        //改变颜色
        msgBt.setTextColor(color_grayLine);
        weekBt.setTextColor(color_blue);

        //开始移动蓝色导航线的 Animation
        movingLineIv.startAnimation(mSlidingRightAnim);
        new Thread(){
            public void run(){
                try {
                    sleep(SLIDE_DURATION);
                } catch (InterruptedException e) {
                    e.printStackTrace();
                }
```

```
                /*
                 *设置蓝色导航线的新位置
                 *将状态改变为一周日程
                 */
                handler.sendEmptyMessage(FLAG_WEEK);
                tabFlag = FLAG_WEEK;
            }
        }.start();
    }
}

private Handler handler = new Handler(){
    @Override
    public void handleMessage(Message msg){
        switch(msg.what){
        case FLAG_WEEK:
            //蓝色导航线的位置
            params.leftMargin = GAP_MOVING_LINE;
            //切换视图
            weekListView.getView().setVisibility(View.VISIBLE);
            msgListView.getView().setVisibility(View.GONE);
            //重新注册功能键(多选或者查询)
            weekListView.registerFunctionButton(funcBt);
            //刷新周视图
            weekListView.refreshView();
            //更新布局
            rootLt.requestLayout();
            break;
        case FLAG_MSG:
            //蓝色导航线的位置
            params.leftMargin = GAP_MOVING_LINE + LENGTH_NAVI_BT;
            //切换视图
            weekListView.getView().setVisibility(View.GONE);
            msgListView.getView().setVisibility(View.VISIBLE);
            //重新注册功能键(多选或者查询)
            msgListView.registerFunctionButton(funcBt);
            //更新布局
            rootLt.requestLayout();
            break;
        }
        //重新设置蓝色导航线的位置
        movingLineIv.clearAnimation();
        movingLineIv.setLayoutParams(params);
        movingLineIv.invalidate();
    }
};
```

其中,weekListView 和 msgListView 这两个对象分别为:

```
private MsgListView msgListView;
private WeekListView weekListView;
```

它们是封装起来的类，分别对应了一周视图（只有一个 ExpandableListView）和过期事件（ExpandableListView 以及按钮）这两个视图。

滑动列表的实现

滑动列表要解决好几个问题。首先，要自己通过 onTouch 事件来计算用户触摸的是哪一行的元素；接着要判断用户是否是真心想要滑动这一行，还是它只是想上下滑动而已；接下来根据触摸的位置滑动某一行，手指离开屏幕，播放 Animation，如果是向左滑动，则删除这一行，让下面的所有行滑动上来；如果是向右滑动，显示删除按钮，直到按这个按钮才真正删除。

首先先获取 ListView 子元素的高度：

```
@Override
public void onDraw(Canvas canvas){
    super.onDraw(canvas);
    View v = FlingExpandListView.this.getChildAt(0);
     //获取 item 的高度
    if(v != null){
        mChildHeight = v.getHeight();
    }
     //初始化网上移动的 Animation
    mSlideUpAnim = new TranslateAnimation(0,0,0, - mChildHeight);
    mSlideUpAnim.setFillAfter(true);
    mSlideUpAnim.setDuration(SLIDE_DURATION);
}
```

这个数据在 ACTION_DOWN 中判断用户触摸的是哪个元素至关重要：

```
@Override
public boolean onTouch(View v,MotionEvent event) {
    //如果此时需要阻塞操作
    if(mIsToBlock){
        return false;
    }
    if(mChildHeight == 0){
        return false;
    }
    int action = event.getAction();
    float x = event.getX();
    float y = event.getY();
    //从第 TOUCH_DOWN 到现在触摸的距离
    float offset = startX - x;
    switch(action){
    case MotionEvent.ACTION_DOWN:
        //记录开始的坐标
        startX = x;
        startY = y;
        //记录开始的时间
```

```
        touchDownTime = System.currentTimeMillis();
        //获取第 1 行的 view
        View view = getChildAt(0);
        if(view == null){
            return false;
        }
        /*
         * 非常重要!!
         * 因为系统会计算每一个 item,不管它有多小
         * 因此,我要先减 1 然后再在 getChildPosition 中加 1
         */
        y -= view.getBottom() - 1;
        mChildPosition = getChildPosition((int) y);

        //如果触摸的是 ListView 的 child,则不滑动
        if(!isGroup(mChildPosition)){
            return skipMovingAction();
        }
```

几个关键函数的定义：

```
/**
 * 获取坐标 y 对应的第几个 item
 * @param y
 * @return
 */
private int getChildPosition(int y){
    if(y <= 0){
        return 0;
    }
    return y/mChildHeight + 1;
}
/**
 * 判断是不是 ExpandableListView 的 child
 * @param position
 * @return
 */
private boolean isGroup(int position){
    if(position == 0){
        return true;
    }
    return !isGroupExpanded(position - 1);
}
/**
 * 跳过 ACTION_MOVE,重置数据
 * @return
 */
private boolean skipMovingAction(){
    mIsToSlide = false;
    timeCount = MAX_TIME;
    return true;
}
```

接下来判断用户是左右滑动的意图还是上下滑动的意图：

```
case MotionEvent.ACTION_MOVE:
    timeCount++;
    //在一小段时间内判断是否左右滑动
    if(timeCount == MAX_TIME){
        if(Math.abs(startY - y) < Math.abs(startX - x)){
            /*
             * 左右的变化量大于上下的变化量,用户正在左右滑动,可以滑动
             * 同时,返回 true,这样就不会再上下滑动
             */
            mIsToSlide = true;
        }else{
            //左右的变化量小于上下的变化量,用户正在上下滑动,不可滑动
            mIsToSlide = false;
        }
    }
```

继续在 ACTION_MOVE 中,如果用户的意图为左右滑动：

```
if(mIsToSlide){
    ((MsgContentAdapter)FlingExpandListView.this.getExpandableListAdapter())
            .onMutiDisappear();
    if(mChildView == null){
        //触摸的是 ListView 下方的空白区域,无操作
        if(mChildPosition >= FlingExpandListView.this.getChildCount()){
            mIsToSlide = false;
            return mIsToSlide;
        }
        //获取子元素的 view
        mChildView = getChildAt(mChildPosition);
    }
    //如果它的 child 打开了,那么就关闭它
    if(FlingExpandListView.this.isGroupExpanded(mChildPosition)){
        FlingExpandListView.this.collapseGroup(mChildPosition);
    }
    /*
     * 标题要长
     * 做了这么多判断,其实滑动的代码就这一句
     */
    mChildView.scrollTo((int) offset,0);
    //正在向左滑动,告诉监听器
    if(offset > 0 && leftSlideEnable){
        if(onItemMovingListener != null){
            onItemMovingListener.onItemMoving(FLAG_MOVING);
        }
    }
}
```

接下来，手放开，这时候开始判断：

```
if(mIsToSlide){
    float speed = offset/(System.currentTimeMillis() - touchDownTime);
    /*
     * 往右滑动
     */
    if(speed < - MAX_SPEED || offset < - MAX_DISTANCE){
        //记录将要删除的一行(也就是出现删除按钮的那一行)
        mChildToRemovePosition = mChildPosition;
        //开始滑动了,屏蔽触摸动作
        blockAciton();
        //往右滑动,删除动作的开始
        slideRightwards(offset);
        flag = FLAG_REMOVE;
    }
    /*
     * 往左滑动
     */
    else if(speed > MAX_SPEED || offset > MAX_DISTANCE){

        //开始滑动了,屏蔽触摸动作
        blockAciton();
        //往左滑动的 Animation 开始
        slideLeftwards(offset);
        flag = FLAG_MOVE;

        if(onItemMovingListener != null){
            onItemMovingListener.onItemMoving(FLAG_MOVED);
        }
    }
    /*
     * 滑回去(见 (2)7)
     */
    else{
        //开始滑动了,屏蔽触摸动作
        blockAciton();
        slideBack(offset);
        //告诉监听器
        if(onItemMovingListener != null){
            onItemMovingListener.onItemMoving(FLAG_CANCELED);
        }
    }
}
```

所谓的 blockAction 也就是将 mIsToBlock 置为 true，这样在 ACTON_DOWN 的时候马上就返回 false 不会有之后的操作了：

```
private void unblockAciton(){
    mIsToBlock = false;
}
private void blockAciton(){
    mIsToBlock = true;
}
```

向左滑动以及向右滑动的代码看起来都差不多，都是开始 Animation，等待 Animation 完成后向 handler 发送消息：

```
private void slideLeftwards(float offset){
    TranslateAnimation slideLeftToAnim = new TranslateAnimation(
            0, - screenWidth + offset,0,0);
    slideLeftToAnim.setFillAfter(true);
    slideLeftToAnim.setDuration(SLIDE_DURATION);
    /*
     * 1.开始 animation
     */
    mChildView.startAnimation(slideLeftToAnim);
    new Thread(){
        public void run(){
            /*
             * 2.等待 animation 的完成
             */
            try {
                sleep(SLIDE_DURATION);
            } catch (InterruptedException e) {
                e.printStackTrace();
            }
            handler.sendEmptyMessage(WHAT_MOVE_DONE);
        }
    }.start();
}
private void slideRightwards(float offset){
    TranslateAnimation slideRightToAnim = new TranslateAnimation(
            0,screenWidth + offset - OFFSET_REMOVE,0,0);
    slideRightToAnim.setFillAfter(true);
    slideRightToAnim.setDuration(SLIDE_DURATION);
    /*
     * 1.开始 animation
     */
    mChildView.startAnimation(slideRightToAnim);
    new Thread(){
        public void run(){
            /*
             * 2.等待 animation 的完成
             */
            try {
                sleep(SLIDE_DURATION);
```

```
            } catch (InterruptedException e) {
                e.printStackTrace();
            }
            /*
             * 3. 显示删除图标(见 4)
             */
            handler.sendEmptyMessage(WHAT_REMOVE_DONE);
        }
    }.start();
}
```

handler 的处理，如果是 WHAT_REMOVE_DONE 那还简单，只要显示删除按钮就可以了：

```
private Handler handler = new Handler(){
    @Override
    public void handleMessage(Message msg){
        switch(msg.what){
        case WHAT_REMOVE_DONE:
            unblockAciton();
            /*
             * 4. 显示删除按钮 (见 5)
             */
            showCloseIcon();
            break;
        case WHAT_MOVE_DONE:
            unblockAciton();
            onListViewBack();
            break;
        }
    }
};
private void showCloseIcon(){
    /*
     * 5. 显示删除按钮
     * 删除动作走了一半
     * 下一步为 6
     * 因为有两种不同的操作(删除或取消),所以接下来会有"(1)" 或 "(2)" 这样的标记,比如 "(1)6"
     */
    mChildView.clearAnimation();
    mChildView.scrollTo(OFFSET_REMOVE - screenWidth,0);
    setChildCloseVisibility(View.VISIBLE);
}
```

但如果是 WHAT_MOVE_DONE，那就要进行下一个 Animation：把下面的元素往上移动，同时等待 Animation 结束后，再发消息给 handler，移动元素：

```
private void onListViewBack(){
    /*
```

```
         * 3.下面的元素开始往上移动
         */
        for(int i = mChildPosition + 1; i < this.getChildCount(); i++){
            View view = this.getChildAt(i);
            view.startAnimation(mSlideUpAnim);
        }
        /*
         * 4.animation 结束后,移动元素(见 5)
         */
        new Thread(){
            public void run(){
                try {
                    sleep(SLIDE_DURATION);
                } catch (InterruptedException e) {
                    e.printStackTrace();
                }
                handler.sendEmptyMessage(WHAT_RETURN);
            }
        }.start();
    }
```

handler 的相应处理以及相关函数:

```
    case WHAT_RETURN:
        unblockAciton();
        /*
         * 5.清除 animation 同时删除 item
         */
        clearAnimationsBelow();
        //告诉监听器
        if(onItemRemoveListener != null){
            onItemRemoveListener.onItemRemove(getChildAbsolutePosition(),flag);
        }
        /*
         * 6.让原来删除的元素回去
         */
        returnViewInDeletedPosition();
        /*
         * 7.搞定
         */
        recycleChildView();
        break;

    //相关函数
    private void clearAnimationsBelow(){
        for(int i = mChildPosition; i < this.getChildCount(); i++){
            View view = this.getChildAt(i);
            view.clearAnimation();
        }
    }
```

```
private void returnViewInDeletedPosition(){
    View viewInDeletedPosition = this.getChildAt(mChildPosition);
    viewInDeletedPosition.scrollTo(0,0);
    viewInDeletedPosition.invalidate();
}
private void recycleChildView(){
    mChildView = null;
}
private int getChildAbsolutePosition(){
    /*
     * 因为给每个位置上的 item 都设置了 id,所以 id 就是绝对位置
     */
    return mChildView.getId();
}
```

需要在 adapter 中设置每个位置上的 item 的 id：

```
@Override
public View getGroupView(final int groupPosition,boolean isExpanded,
        View convertView,ViewGroup parent) {
    //省略代码
    convertView.setId(groupPosition);
    return convertView;
}
```

到此为止，移动的动作完成了，但是删除的动作才完成一半。接下来，如果有元素已经进入"待删除"状态(显示了删除按钮)，那么就要进行下一轮的判断。

在 ACTION_DOWN 中：

```
//(1)6  如果有等待删除的 item 且用户正在单击删除按钮,跳过 ACTION_MOVE
if(mChildToRemovePosition == mChildPosition){
    return skipMovingAction();
}
/*
 * (2)6 如果有等待删除的 item 但是用户没有单击删除按钮,跳过 ACTION_MOVE
 *       同时滑动回去(见 (2)7)
 */
else if(mChildToRemovePosition != NULL_POSITION){
    slideBack(offset);
    return skipMovingAction();
}
```

滑回去的操作跟之前的大同小异，也是播放 Animation，然后让 handler 善后：

```
private void slideBack(float offset){
    /*
     * (2)7 开始 animation 同时删除按钮消失
     */
    TranslateAnimation slideLeftBackAnim = new TranslateAnimation(0,offset,0,0);
```

```
        slideLeftBackAnim.setFillAfter(true);
        slideLeftBackAnim.setDuration(SLIDE_DURATION);
        setChildCloseVisibility(View.GONE);
        mChildView.startAnimation(slideLeftBackAnim);
        new Thread(){
            public void run(){
                try {
                    sleep(SLIDE_DURATION);
                } catch (InterruptedException e) {
                    e.printStackTrace();
                }
                /*
                 * (2)8 滑回来(见 (2)9)
                 */
                handler.sendEmptyMessage(WHAT_REMOVE_BACK);
            }
        }.start();
    }
    private void setChildCloseVisibility(int v){
        ImageView closeIv = (ImageView)
              mChildView.findViewById(R.id.list_msg_close_iv);
        closeIv.setVisibility(v);
    }
```

handler 的代码：

```
case WHAT_REMOVE_BACK:
    unblockAciton();
    /*
     * (2)9 所有的 item 都返回
     */
    putChildViewBack();
    /*
     * (2)10 回收 child
     */
    recycleChildView();
break;
```

接下来，在 ACTION_UP 中判断是不是按了这个删除按钮：

```
/*
 * (1)7  按下删除按钮
 */
if(mChildToRemovePosition == mChildPosition){
    if(x > screenWidth - OFFSET_REMOVE){
        /*
         * (1)7(1) 按下删除按钮
         * 先隐藏这一行以及删除按钮(实际上隐藏的是下面一行的删除按钮)
         */
        mChildView.scrollTo(-screenWidth,0);
```

```
        setChildCloseVisibility(View.GONE);
        /*
         *(1)8  让下面的行向上滑动(参照 (1)9)
         */
        onListViewBack();
    }else{
        /*
         *(1)7(2) 单击空白区域,滑回来
         *跟 (2)6 一样
         */
        slideBack(OFFSET_REMOVE - screenWidth);
    }
    recycleChildToRemove();
    reset();
    return true;
}
/*
 *(1)7*没有单击删除按钮
 */
else if(mChildToRemovePosition != NULL_POSITION){
    recycleChildToRemove();
    reset();
    return true;
}

//相关函数
private void recycleChildToRemove(){
    mChildToRemovePosition = NULL_POSITION;
}
private void reset(){
    mIsToSlide = false;
    timeCount = 0;
}
```

一周日程的实现

一周日程使用了常规的 ExpandableListView,但又不是那么常规,因为当 collapse 的时候,它的视图如图 4-14 所示。

而当一个列表打开时,列表如图 4-15 所示。

图 4-14 列表关闭　　图 4-15 列表打开

这样的效果跟前面切换一周日程和过期事件的过程类似,也是通过重叠设置两个 View,然后切换设置 visibility 的方式来进行的。

对应的 layout 为：

```
<!--用于展开后显示日程的布局,关闭时为 invisible-->
<RelativeLayout
    android:id="@+id/event_parent_content_lt"
    android:layout_width="wrap_content"
    android:layout_height="wrap_content"
    android:visibility="invisible"
    android:layout_marginLeft="8dip"
    android:layout_toLeftOf="@+id/event_parent_expand_iv"
    android:layout_toRightOf="@+id/line">

    <TextView
        android:id="@+id/event_parent_time_tv"
        android:layout_width="wrap_content"
        android:layout_height="wrap_content"
        android:layout_marginLeft="8dip"
        android:text="@string/week_time"
        android:textColor="@color/gray"
        android:textSize="35dip"/>

    <TextView
        android:id="@+id/event_parent_content_tv"
        android:layout_width="wrap_content"
        android:layout_height="wrap_content"
        android:layout_marginLeft="2dip"
        android:layout_alignLeft="@+id/event_parent_time_tv"
        android:layout_below="@+id/event_parent_time_tv"
        android:text="@string/week_content"
        android:maxLines="1"
        android:textColor="@color/gray"
        android:textSize="17dip"/>
</RelativeLayout>

<!--用于展开后显示日程的布局-->
<LinearLayout
    android:id="@+id/event_parent_button_lt"
    android:layout_width="wrap_content"
    android:layout_height="wrap_content"
    android:layout_alignBottom="@+id/event_parent_content_lt"
    android:layout_alignRight="@+id/event_parent_content_lt"
    android:layout_alignParentTop="true"
    android:layout_marginLeft="8dip"
    android:layout_toRightOf="@+id/line"
    android:gravity="right|bottom">
    <!--个数图片-->
    <ImageView
        android:id="@+id/event_parent_figure_iv"
        android:focusable="false"
        android:layout_width="wrap_content"
```

```
        android:layout_height = "wrap_content"
        android:background = "@drawable/ic_0"
    />
    <!-- 增加按钮 -->
    < ImageButton
        android:id = "@+ id/event_parent_add_iv"
        android:focusable = "false"
        android:layout_width = "wrap_content"
        android:layout_height = "wrap_content"
        android:background = "@drawable/ic_add"
    />
</LinearLayout >
```

在 adapter 中添加相应的代码。由于 ListView 回收 View 特性的缘故，我们必须得使用一个额外的数据来存放已打开的行：

```
private HashSet < Integer > expandedGroup;
```

并通过按钮的单击事件来加入到这个 HashSet 中：

```
@Override
public View getGroupView(final int groupPosition,boolean isExpanded,
        View convertView,ViewGroup parent) {
    Log.e("getGroupView",groupPosition + "");
    final GroupViewHolder holder;
    if(convertView == null){
        //省略代码
    }else{
        holder = (GroupViewHolder)convertView.getTag();
    }
    //设置按钮监听器
    setButtonsListener(holder,groupPosition);
    //设置是否展开列表
    setExpandedGroup(holder,groupPosition);

    //又省略代码
}
private void setButtonsListener(GroupViewHolder holder,final int groupPosition){

    OnClickListener clickListener = new OnClickListener(){
        @Override
        public void onClick(View v) {
            WeekTimeVo vo = (WeekTimeVo) group.get(groupPosition);
            if(vo.getEvents().size() == 0){
                return;
            }
            if(listView.isGroupExpanded(groupPosition)){
                expandedGroup.remove(groupPosition);
```

```
            }else{
                expandedGroup.add(groupPosition);
            }
            listView.invalidateViews();
        }
    };
    /**
     * 单击指示器打开/关闭
     */
    holder.expand.setOnClickListener(clickListener);
    /**
     * 单击空白打开/关闭
     */
    holder.buttonLt.setOnClickListener(clickListener);
     //继续省略代码
}
```

再判断某一行标记为打开或者关闭：

```
private void setExpandedGroup(GroupViewHolder holder,final int groupPosition){
    //如果已经打开了就关闭
    if(expandedGroup.contains(groupPosition)){
        holder.contentLt.setVisibility(View.VISIBLE);
        holder.buttonLt.setVisibility(View.INVISIBLE);
        holder.expand.setBackgroundDrawable(
                res.getDrawable(R.drawable.ic_collapse));
        listView.expandGroup(groupPosition);
    }
    //如果关闭了就打开
    else{
        holder.contentLt.setVisibility(View.INVISIBLE);
        holder.buttonLt.setVisibility(View.VISIBLE);
        holder.expand.setBackgroundDrawable(
               res.getDrawable(R.drawable.ic_expand));
        listView.collapseGroup(groupPosition);
    }
}
```

第五章　旋转控件

用 SurfaceView 自制超炫控件

清单

Demo 代码：

```
\demo\Demo_SurfaceView
\demo\Demo_FloatView
```

实例代码：

```
\source_codes\SpinWidget
\source_codes\Clock
```

Target SDK：

```
Android 2.1
```

第一节　产 品 介 绍

Android 自带的控件基本上可以满足我们的使用。但是在某些情况下为了提升用户体验，我们需要自己设计一个特定的控件。在正常情况下，控件缩放于左下角。单击按钮后，菜单按钮会全部弹出，如图 5-1 所示。

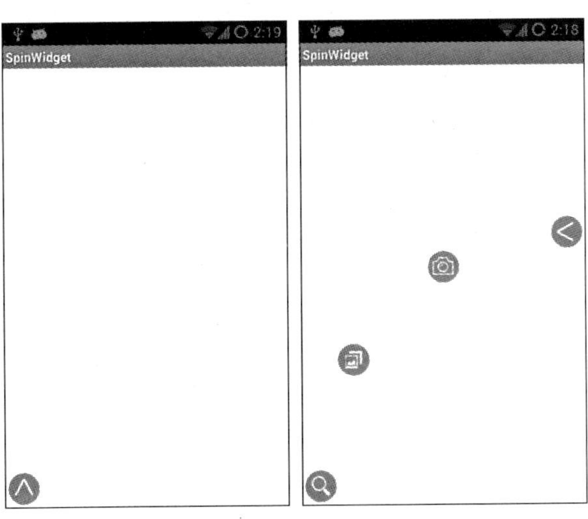

图 5-1　菜单

第二节 画图专用 SurfaceView

Android 的游戏开发，几乎都是使用 SurfaceView。相比于 View，SurfaceView 的优点在于速度快，速度快，速度快……总之就是速度快。SurfaceView 一般包括几个部分：用于控制绘图的线程，触摸事件，以及初始化创建函数。它的一般框架为：

```
public class ViewFrame extends SurfaceView implements
            SurfaceHolder.Callback,View.OnTouchListener,Runnable{

    public ViewFrame(Context context,AttributeSet attrs) {
        super(context,attrs);
        //构造函数
    }

    @Override
    public boolean onTouch(View v,MotionEvent event) {
        //触摸事件
        return false;
    }

    @Override
    public void surfaceCreated(SurfaceHolder holder) {
        //SurfaceView 被创建的初始化操作
    }

    @Override
    public void surfaceChanged(SurfaceHolder holder,int format,int width,
            int height) {
        //SurfaceView 被改变的操作
    }

    @Override
    public void surfaceDestroyed(SurfaceHolder holder) {
        //SurfaceView 被销毁的初始化操作
    }

    @Override
    public void run() {
        //画图线程
    }

}
```

使用 SurfaceView 的过程一般如下。

(1) 获取 SurfaceHolder 并添加回调函数：

SurfaceHolder 在 SurfaceView 中很重要，它是 SurfaceView 的控制器，通过 Surfaceview.

getHolder()来获得。使用 SurfaceView 有一个原则，所有的绘图工作必须得在 Surface 被创建之后才能开始，而在 Surface 被销毁之前必须结束。SurfaceCreated 和 SurfaceDestroyed 就形成绘图处理的边界。在这个区间以外绘图就会报空指针异常，这是所有新学者都会遇到的问题。一般来说，在使用 SurfaceHolder. lockCanvas()这个方法返回空都是这个原因。

```
//添加回调函数
SurfaceHolder holder = this.getHolder();
holder.addCallback(this);
```

(2) 获取画布、绘图、提交画布：

```
//获取画布,锁定画布
Canvas canvas = holder.lockCanvas();
/*
 * 在画布上绘图
 * 绘图的动作要在 SurfaceHolder.lockCanvas()和
    SurfaceHolder.unlockCanvasAndPost(canvas)之间进行
 */
canvas.drawBitmap(bm,0,0,null);
//提交画布
holder.unlockCanvasAndPost(canvas);
```

最简单的 SurfaceView

我们从最简单的例子开始。这是一个绘制一个图形的 SurfaceView。

```
public class BasicView extends SurfaceView implements SurfaceHolder.Callback{

    //待绘制的位图
    private Bitmap bm;

    public BasicView(Context context,AttributeSet attrs) {
        super(context,attrs);

        //获取位图
        bm = BitmapFactory.decodeResource(context.getResources(),
                R.drawable.ic_launcher);
        //添加回调函数
        SurfaceHolder holder = this.getHolder();
        holder.addCallback(this);

         //设置为窗体的最高层
        setZOrderOnTop(true);
        //设置透明格式
        holder.setFormat(PixelFormat.TRANSLUCENT);
    }

    @Override
```

```
    public void surfaceCreated(SurfaceHolder holder) {
        //回调函数,当 SurfaceView 初始化时调用这个函数,绘制图像

        //获取画布,锁定画布
        Canvas canvas = holder.lockCanvas();
        /*
         * 在画布上绘图
         * 绘图的动作要在 SurfaceHolder.lockCanvas()和
         * SurfaceHolder.unlockCanvasAndPost(canvas)之间进行
         */
        canvas.drawBitmap(bm,0,0,null);
        //提交画布
        holder.unlockCanvasAndPost(canvas);
    }

    @Override
    public void surfaceChanged(SurfaceHolder holder, int format, int width,
            int height) {
    }

    @Override
    public void surfaceDestroyed(SurfaceHolder holder) {
    }
}
```

Activity 的代码非常简单,只指定了布局文件:

```
public class BasicActivity extends Activity {

    @Override
    public void onCreate(Bundle savedInstanceState) {
        super.onCreate(savedInstanceState);
        setContentView(R.layout.activity_basic);
    }
}
```

布局文件:

```
<RelativeLayout xmlns:android="http://schemas.android.com/apk/res/android"
    xmlns:tools="http://schemas.android.com/tools"
    android:layout_width="fill_parent"
    android:layout_height="fill_parent" >

    <com.example.demo_surfaceview.view.BasicView
        android:layout_width="fill_parent"
        android:layout_height="fill_parent"/>

</RelativeLayout>
```

在接下来的几节中,Activity 的代码都是设定了相应的布局文件,布局文件也都是设定相应的控件,因此不再赘述。

SurfaceView 绘图机制

为了理解好 SurfaceView 的绘图机制，我做了一个小例子，如图 5-2 所示。

图 5-2　SurfaceView 的一个测试实例

每次单击按钮，SurfaceView 都会调用一次绘图的方法：

```
public class TestActivity extends Activity {

    private TestSurfaceView view;
    private int i = 1;

    @Override
    public void onCreate(Bundle savedInstanceState) {
        super.onCreate(savedInstanceState);
    setContentView(R.layout.activity_test);

    view = (TestSurfaceView) findViewById(R.id.test);
    }

    //按下按钮时的事件
    public void click(View v){
        view.draw(i++);
    }
}
```

这张 SurfaceView 的代码如下：

```
public class TestSurfaceView extends BasicView {

    //画笔
```

```
        private Paint paint = new Paint();

        public TestSurfaceView(Context context,AttributeSet attrs) {
            super(context,attrs);
            paint.setTextSize(50);
            paint.setColor(context.getResources().getColor(R.color.blue));
        }

        @Override
        public void surfaceCreated(SurfaceHolder holder) {
            draw(0);
        }

        /**
         * 绘制数字
         * @param i 数字
         */
        public void draw(int i){
            draw((i + 2) * 20,(i + 2) * 20,i);
        }

        private void draw(float x,float y,int i){
            Canvas canvas = mHolder.lockCanvas();
            //绘制数字
            canvas.drawText(i + "",x,y,paint);
            mHolder.unlockCanvasAndPost(canvas);
        }
    }
```

按照常理来说，第 0 次单击按钮(surfaceCreated)时，画布上画出 0；第 1 次单击按钮时，由于没有清空原来画布的内容，因此画布上会画出 0 和 1，第 2 次画出 0、1、2、…以此类推。

但实际上并没有达到预期的效果，如图 5-3 所示。

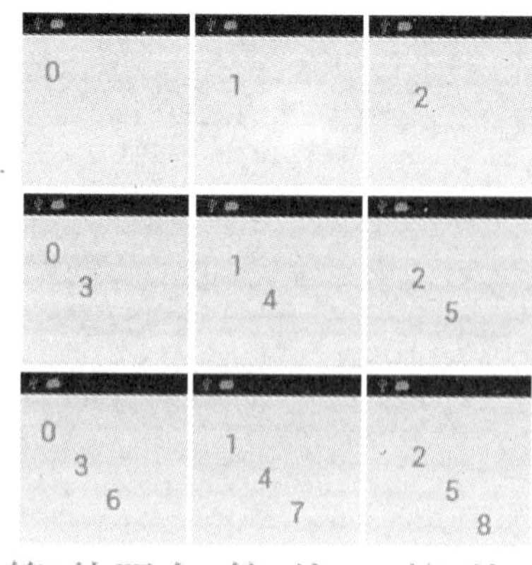

图 5-3　SurfaceView 测试结果

这是因为 SurfaceView 在绘图的时候使用了三块画布轮流绘图的机制(在早期的 Android 版本中,使用的是两块画布)。这样做的目的是通过牺牲内存来提高显示性能。

第三节 OnTouchListener 详解

初探 OnTouchListener

View 这个类一个重要的监听器就是 OnTouchListener,它能监听所有在这个 View 上的触摸事件,包括触摸坐标、行为、触摸点数甚至触摸压力。一个 TouchListener 事件一般都会需要这样几个信息:x 坐标,y 坐标,行为(第一次触摸、移动、松开),如:

```
View view = (View) findViewById(R.id.touch);
view.setOnTouchListener(new OnTouchListener() {
    @Override
    public boolean onTouch(View view,MotionEvent event) {
        //触摸事件发生的 x 相对坐标
        float x = event.getX();
        //触摸事件发生的 x 绝对坐标(相对于整个屏幕)
        float rawX = event.getRawX();
        //触摸事件发生的 y 相对坐标
        float y = event.getY();
        //触摸事件发生的 y 绝对坐标(相对于整个屏幕)
        float rawY = event.getRawY();
        //获取事件行为
        int action = event.getAction();
        switch(action){
        //手指刚触摸屏幕
        case MotionEvent.ACTION_DOWN:
            break;
        //手指松开
        case MotionEvent.ACTION_UP:
            break;
        //移动
        case MotionEvent.ACTION_MOVE:
            break;
        }
        /*
         * 当返回 false 时,将不会接收 ACTION_MOVE 跟 ACTION_UP 事件
         */
        return true;
        }
});
```

坐标 x,y 跟坐标 rawX,rawY 的区别在于,前者是在 view 上面的坐标,而后者是在整个界面上的坐标。如图 5-4 所示,触摸图中图标时,x 与 y 的值为 3,14,而 rawX 与 rawY 的值为 195,385。

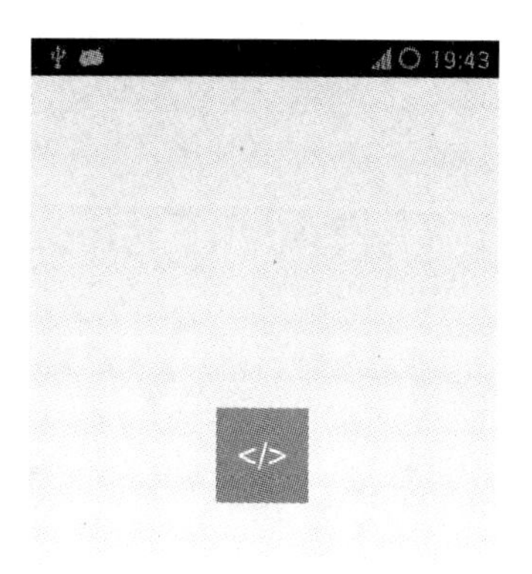

图 5-4 OnTouchListener 的一个例子

实例：触摸绘图

接下来我们来实现图形随着手指的移动而移动的例子。

```
public class TouchView extends BasicView implements View.OnTouchListener{

    private Rect imgRect = new Rect();
    private boolean isToDraw = false;
    private float offsetX;
    private float offsetY;

    public TouchView(Context context,AttributeSet attrs) {
        super(context,attrs);
        //添加触摸事件
        setOnTouchListener(this);
    }

    @Override
    public void surfaceCreated(SurfaceHolder holder) {
        draw(0,0);
    }

    protected void draw(float x,float y){
        saveRect((int)x,(int)y);
        Canvas canvas = mHolder.lockCanvas();
        //清空原来的图像
        canvas.drawColor(Color.TRANSPARENT,Mode.CLEAR);
        canvas.drawBitmap(bm,x,y,null);
        mHolder.unlockCanvasAndPost(canvas);
    }
```

```
    /**
     * 保存图形的位置
     * @param x
     * @param y
     */
    protected void saveRect(int x,int y){
        imgRect.left = x;
        imgRect.top = y;
        imgRect.right = x + bm.getWidth();
        imgRect.bottom = y + bm.getHeight();
    }

    @Override
    public boolean onTouch(View v,MotionEvent event) {
        float x = event.getX();
        float y = event.getY();
        switch(event.getAction()){
        case MotionEvent.ACTION_DOWN:
            //触摸点是否落在图形之内
            if(imgRect.contains((int)x,(int)y)){
                offsetX = x - imgRect.left;
                offsetY = y - imgRect.top;
                isToDraw = true;
            }
            break;
        case MotionEvent.ACTION_MOVE:
            if(isToDraw){
                draw(x - offsetX,y - offsetY);
            }
            break;
        case MotionEvent.ACTION_UP:
            isToDraw = false;
            break;
        }
        return true;
    }
}
```

区域绘图

在绘制图形的时候，对特定的区域进行绘图，不需要对整个区域刷新。

```
public class EnhancedTouchView extends TouchView{

    public EnhancedTouchView(Context context,AttributeSet attrs) {
        super(context,attrs);
    }

    private float beforeX;
    private float beforeY;
```

```
    @Override
    protected void draw(float x,float y){
        saveRect((int)x,(int)y);

        //按区域进行刷新
        Rect rect = getDrawingRect(x,y);
        Canvas canvas = mHolder.lockCanvas(rect);

        canvas.drawColor(Color.TRANSPARENT,Mode.CLEAR);
        canvas.drawBitmap(bm,x,y,null);
        mHolder.unlockCanvasAndPost(canvas);

        //保存之前的绘图坐标
        beforeX = x;
        beforeY = y;
    }

    /**
     * 获取需要绘图的 Rect
     * @param x
     * @param y
     * @return
     */
    private Rect getDrawingRect(float x,float y){
        Rect rect = new Rect();
        rect.left = (int) Math.min(beforeX,x);
        rect.top = (int) Math.min(beforeY,y);
        rect.right = (int) Math.max(beforeX,x) + bm.getWidth();
        rect.bottom = (int) Math.max(beforeY,y) + bm.getHeight();
        return rect;
    }

}
```

轨迹绘图

接下来实现的功能，让图标跟随手指移动的同时，沿着一个特定轨迹移动。这个特定轨迹为以屏幕正中心为中心点的圆。

因此我们在进行坐标的计算时要进行多次转换：

(1) 正常坐标系：绝对坐标 AbsolutePoint。

(2) 以屏幕中心为原点的坐标系：相对坐标 RelativePoint。

(3) 绘图坐标：ResultPoint。

这些坐标的计算操作都封装在 SinglePointManager 中。SinglePointManager 的全局变量和构造函数为：

```
//圆周半径
protected float r;
//圆心坐标
```

```
protected float rx;
protected float ry;
//图标宽度
protected float iconWidth;

/**
 * @param r 圆的半径
 * @param rx 圆心坐标 x
 * @param ry 圆心坐标 y
 * @param iconWidth 图片的宽度
 */
public SinglePointManager(float r,float rx,float ry,float iconWidth)
{
    this.r = r;
    this.rx = rx;
    this.ry = ry;
    this.iconWidth = iconWidth;
}
```

如图 5-5 所示，在图中有三个点 A、B、C，其中 A 是手指的位置，C 是图形的绘图位置（图形的左上角），坐标中心的数值表示 AC 与 y 轴正半轴的角度。

为了得到点 C 的确切位置，要经过以下几个步骤。

(1) 首先，由于系统的坐标系以屏幕左上角为中心，所以要先对坐标系进行转换，将坐标系转换成以中心为圆心的坐标系，以便计算，如图 5-6 所示。

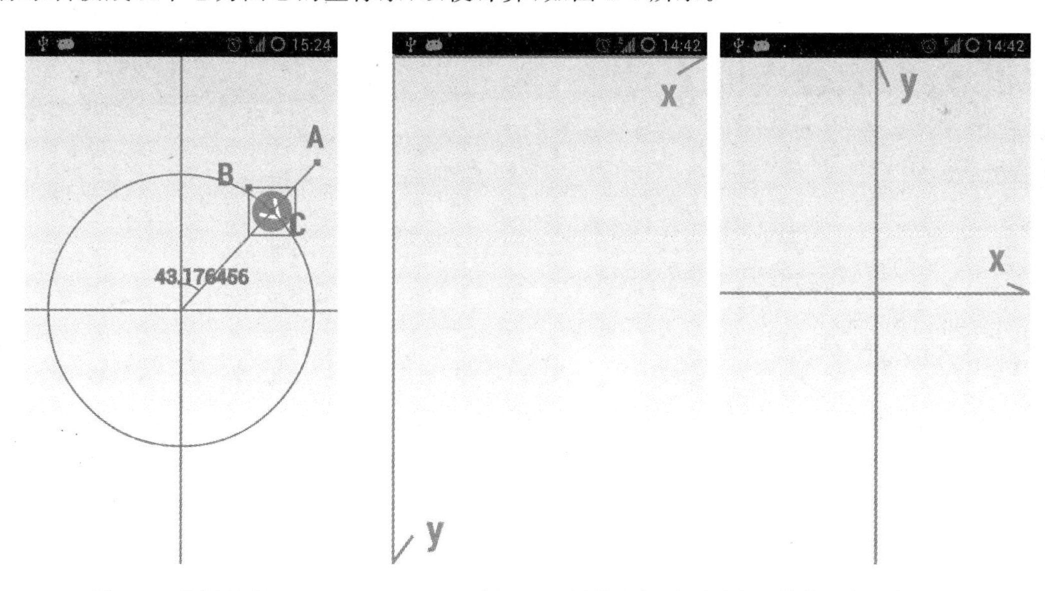

图 5-5　绘图示意

图 5-6　原始坐标系(左图)和转换后的坐标系

```
Point relativePoint = getRelativePointByAbsolutePoint(p);
/**
 * @param p 绝对坐标
```

```
 * @return 相对坐标
  */
protected Point getRelativePointByAbsolutePoint(Point p)
{
    return new Point(p.getX() - rx,ry - p.getY());
}
```

(2) 计算 AC 与 y 轴正半轴的角度：

```
//根据绝对坐标获取在相对坐标下与 y 轴正半轴的夹角
float angle = getAngleByRelativePoint(relativePoint);
//当角度为 2PI 时,取零
if(Math.abs(angle - 2 * Math.PI) <= 0.01){
    angle = 0;
}
/**
  * 获取在相对坐标系下该点与圆心连线和 y 轴正半轴所成的角度
  * @param pt 相对坐标
  * @return
  */
protected float getAngleByRelativePoint(Point pt){
    float angle = (float) Math.abs(Math.atan(pt.getX()/pt.getY()));
    //获取坐标的象限
    switch(getQuadrant(pt)){
    case 4:
        angle = (float) (Math.PI - angle);
        break;
    case 3:
        angle += Math.PI;
        break;
    case 2:
        angle = (float) (Math.PI * 2 - angle);
        break;
    }
    return angle;
}
```

(3) 计算点 C 在相对坐标系中的坐标：

```
relativePoint = new Point(FloatMath.sin(angle) * r,FloatMath.cos(angle) * r);
```

(4) 转换坐标系：

```
Point absolutePoint = getAbsolutePointByRelativePoint(pointInCenterAxis);
/**
  *
  * @param p 相对坐标
  * @return 绝对坐标
  */
```

```
protected Point getAbsolutePointByRelativePoint(Point p)
{
    return new Point(p.getX() + rx,ry - p.getY());
}
```

(5) 返回点 B 的位置：

```
Point resultPoint = getIconPoint(absolutePoint,true);
/**
 * @param p
 * @param isAbsolute 是相对坐标还是绝对坐标
 * @return 图标的绘图坐标
 */
protected Point getIconPoint(Point p,boolean isAbsolute)
{
    float retX = 0;
    float retY = 0;
    if(isAbsolute){
        retX = p.getX() - iconWidth/2;
        retY = p.getY() - iconWidth/2;
    }else{
        retX = p.getX() - iconWidth/2;
        retY = p.getY() + iconWidth/2;
    }
    return new Point(retX,retY);
}
```

因此,我们只要再调用一个方法就可以将触摸的坐标转换成绘图坐标。

```
/**
 * @param p 绝对坐标下的触摸坐标
 * @return 绘图坐标
 */
public Point getResultPointByAbsolutePoint(Point p){
    //将绝对坐标转换成相对坐标
    Point relativePoint = getRelativePointByAbsolutePoint(p);
    //根据绝对坐标获取在相对坐标下与 y 轴正半轴的夹角
    float angle = getAngleByRelativePoint(relativePoint);
    //当角度为 2PI 时,取零
    if(Math.abs(angle - 2 * Math.PI) <= 0.01){
        angle = 0;
    }
    //计算相对坐标系下的坐标
    relativePoint = new Point(FloatMath.sin(angle) * r,FloatMath.cos(angle) * r);
    //获取绝对坐标系下的坐标
    Point absolutePoint = getAbsolutePointByRelativePoint(relativePoint);
    //获取图标的绘图坐标(左上角)
    Point resultPoint = getIconPoint(absolutePoint,true);
    return resultPoint;
}
```

这个 View 的代码如下：

```
public class SpinView extends EnhancedTouchView{

    protected SinglePointManager pm;
    private boolean mIsLoaded = false;

    //屏幕的长宽
    private int screenWidth;
    private int screenHeight;
    //图标的宽度
    protected int iconWidth;
    //半径
    private int r;
    //圆心坐标
    protected int rx;
    protected int ry;

    protected int color_blue;
    protected int color_red;
    private Paint paint = new Paint();

    public SpinView(Context context,AttributeSet attrs) {
        super(context,attrs);

        Resources res = context.getResources();
        color_blue = res.getColor(R.color.blue);
        color_red = res.getColor(R.color.red);
        paint.setColor(color_blue);
        paint.setStrokeWidth(3);
        paint.setStyle(Paint.Style.STROKE);
        /*
         * 设置 View 加载完成的监听器，当 View 加载完成时长宽不为 0
         */
        getViewTreeObserver().addOnGlobalLayoutListener(
                    new OnGlobalLayoutListener() {
            @Override
            public void onGlobalLayout() {
                if(!mIsLoaded){
                    initPositionManager();
                    mIsLoaded = true;
                }
            }
        });
    }

    private void initPositionManager(){
        screenWidth = this.getWidth();
        screenHeight = this.getHeight();
        iconWidth = bm.getWidth();
```

```
        r = screenWidth/2 - iconWidth/2;
        rx = screenWidth/2;
        ry = screenHeight/2;

        pm = new SinglePointManager(r,rx,ry,iconWidth);
    }

    @Override
    public void surfaceCreated(SurfaceHolder holder) {
        drawByAngle(0);
    }

    @Override
    protected void draw(float x,float y){
        Canvas canvas = mHolder.lockCanvas();
        canvas.drawColor(Color.TRANSPARENT,Mode.CLEAR);

        /*
         * 关键,获取坐标函数
         */
        Point pt = pm.getResultPointByAbsolutePoint(new Point(x,y));
        saveRect((int)pt.getX(),(int)pt.getY());
        canvas.drawBitmap(bm,pt.getX(),pt.getY(),null);

        /*
         * 绘制辅助图形
         * 可以删掉
         */
        drawUtils(canvas,x,y);

        mHolder.unlockCanvasAndPost(canvas);

    }

    protected void drawUtils(Canvas canvas,float x,float y){
        paint.setColor(color_blue);
        paint.setStrokeWidth(3);
        //绘制坐标
        canvas.drawLine(0,ry,screenWidth,ry,paint);
        canvas.drawLine(rx,0,rx,screenHeight,paint);
        //绘制圆圈
        canvas.drawCircle(rx,ry,r,paint);

        //从原点到手指触摸点的线段
        canvas.drawLine(rx,ry,x,y,paint);

        //绘制正方形区域
        Point resultPt = pm.getResultPointByAbsolutePoint(new Point(x,y));
        canvas.drawRect(resultPt.getX(),resultPt.getY(),
                resultPt.getX() + iconWidth,resultPt.getY() + iconWidth,paint);
```

```
        paint.setColor(color_red);
        paint.setStrokeWidth(10);

        //绘制手指触摸点
        canvas.drawPoint(x,y,paint);

        //绘制在圆圈上的点
        Point roundPt = pm.testGetPointAroundTrack(new Point(x,y));
        canvas.drawPoint(roundPt.getX(),roundPt.getY(),paint);

        //绘制绘图点
        canvas.drawPoint(resultPt.getX(),resultPt.getY(),paint);

    }

    //根据角度绘制图形
    protected void drawByAngle(float angle){
        Point pt = pm.getAbsolutePointByAngle(angle);
        draw(pt.getX(),pt.getY());
    }

    /*
     * 去掉 offsetX 和 offsetY
     */
    @Override
    public boolean onTouch(View v,MotionEvent event) {
        float x = event.getX();
        float y = event.getY();
        switch(event.getAction()){
        case MotionEvent.ACTION_DOWN:
            if(imgRect.contains((int)x,(int)y)){
                isToDraw = true;
            }
            break;
        case MotionEvent.ACTION_MOVE:
            if(isToDraw){
                draw(x,y);
            }
            break;
        case MotionEvent.ACTION_UP:
            isToDraw = false;
            break;
        }
        return true;
    }
}
```

此外,在这个控件中提供了根据角度绘图的函数,这个函数很简短:

```
//根据角度绘制图形
protected void drawByAngle(float angle){
    Point pt = pm.getAbsolutePointByAngle(angle);
    draw(pt.getX(),pt.getY());
}
```

函数 getAbsolutePointByAngle 获取的是在绝对坐标下的图标的中心坐标(注意,是中心坐标),然后把这个中心坐标丢给 draw 函数去处理。这个函数将会在接下来的 RollBackView 中起到重要作用。

```
/*
 * @param pt 绝对坐标
 * @return 在相对坐标系下与 y 轴正半轴的夹角
 */
public float getAngleByAbsolutePoint(Point pt){
    Point p = getRelativePointByAbsolutePoint(pt);
    return getAngleByRelativePoint(p);
}
```

第四节　图形变换 Matrix

旋转绘图

在 SpinView 的基础上,我们让图标随着角度的变化而旋转,这时需要获取旋转角度,如图 5-7 所示。

```
/*
 * @param pt 绝对坐标
 * @return 在相对坐标系下与 y 轴正半轴的夹角
 */
public float getAngleByAbsolutePoint(Point pt){
    Point p = getRelativePointByAbsolutePoint(pt);
    return getAngleByRelativePoint(p);
}
```

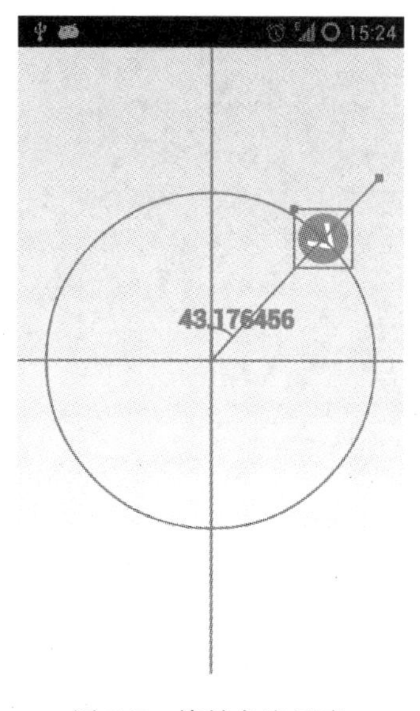

图 5-7　旋转角度示意

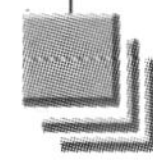

```
public class MatrixView extends SpinView{

    private Matrix matrix = new Matrix();

    public MatrixView(Context context, AttributeSet attrs) {
        super(context, attrs);

        //设置抗锯齿
        paint.setAntiAlias(true);
    }

    @Override
    protected void draw(float x, float y){
        Point originalPt = new Point(x, y);
        Point pt = pm.getResultPointByAbsolutePoint(originalPt);
        saveRect((int)pt.getX(), (int)pt.getY());

        //图形的旋转角度
        float angle = pm.getAngleByAbsolutePoint(originalPt);
        angle = (float) (180 * angle/Math.PI);

        Canvas canvas = mHolder.lockCanvas();
        canvas.drawColor(Color.TRANSPARENT, Mode.CLEAR);

        //旋转
        matrix.reset();
        matrix.setTranslate(pt.getX(), pt.getY());
        matrix.preRotate(angle, (float)bm.getWidth()/2, (float)bm.getWidth()/2);
        canvas.drawBitmap(bm, matrix, paint);

        drawUtils(canvas, x, y, angle);

        mHolder.unlockCanvasAndPost(canvas);
    }

    protected void drawUtils(Canvas canvas, float x, float y, float angle){
        super.drawUtils(canvas, x, y);

        //在坐标圆心中绘制弧形
        paint.setColor(color_blue);
        paint.setStrokeWidth(3);
        canvas.drawArc(new RectF(rx - 40, ry - 40, rx + 40, ry + 40), - 90, angle, true, paint);

        //绘制角度数字
        paint.setColor(color_red);
        paint.setTextSize(30);
        canvas.drawText(angle + "", rx - 40, ry - 40, paint);
    }
}
```

接下来，我们要实现当手指离开屏幕时，图标会按照原来的轨迹滑回原点。这时候需要使用一个线程来控制这个过程。在这个线程中，使用 drawByAngle(float angle)函数来绘制图形，通过不断增加（减少）angle 的值来达到动态的效果：

```
public class RollBackView extends MatrixView{

    private boolean mIsBlocked = false;

    public RollBackView(Context context,AttributeSet attrs) {
        super(context,attrs);
    }

    @Override
    public boolean onTouch(View v,MotionEvent event) {
        if(mIsBlocked){
            return false;
        }
        float x = event.getX();
        float y = event.getY();
        switch(event.getAction()){
        case MotionEvent.ACTION_DOWN:
            if(imgRect.contains((int)x,(int)y)){
                isToDraw = true;
            }
            break;
        case MotionEvent.ACTION_MOVE:
            if(isToDraw){
                draw(x,y);
            }
            break;
        case MotionEvent.ACTION_UP:
            //开始移动
            new RollbackThread(pm.getAngleByAbsolutePoint(new Point(x,y)),0.1f,0).start();
            isToDraw = false;
            break;
        }
        return true;
    }

    class RollbackThread extends Thread{

        //终点的角度
        private float destination = 0;
        //方向
        private int direction = 1;
        //运行控制
        private boolean isRun = true;
        //移动时的角度
        private float angle;
        //移动速度
        private float movingOffset;
```

```
        public RollbackThread(float angle,float movingOffset,float destination){
            this.angle = angle;
            this.movingOffset = movingOffset;
            this.destination = destination;
            if(angle < Math.PI){
                direction = -1;
            }
        }

        public void run(){
            mIsBlocked = true;
            while(isRun){
                  if (Math.abs (angle - destination) <= movingOffset || Math.abs (angle - destination) >= Math.PI * 2){
                    isRun = false;
                    drawByAngle(destination);
                }else{
                    angle += direction * movingOffset;
                    drawByAngle(angle);
                }
            }
            mIsBlocked = false;
        }
    }
}
```

图标组的移动

接下来我们来制作一个更为复杂的控件，拖动一个图标，其他图标跟着移动，并且保持相同的距离，如图 5-8 所示。它的关键在于获取第 0 个图标与 y 轴正半轴的角度，其余的图标与 y 轴正半轴的夹角就可以根据这个夹角计算出来。

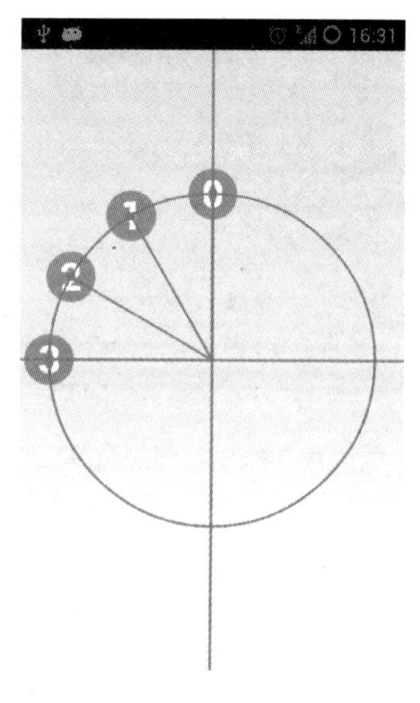

图 5-8　图标组

ImageSetView 继承于 RollBackView：

```
public class ImageSetView extends RollBackView{

    //绘制的位图
    protected Bitmap[] icons = new Bitmap[4];

    //继承于 SinglePointManager
    protected MutiPointManager mPm;

    //触摸的图标序号
    protected int icon = 0;

    //开始和结束的角度
    protected static final float START_ANGLE = (float) Math.PI * 2;
    protected static final float END_ANGLE = (float) Math.PI * 3/2;

    public ImageSetView(Context context,AttributeSet attrs) {
        super(context,attrs);

        Resources res = context.getResources();
        icons[0] = BitmapFactory.decodeResource(res,R.drawable.ic_0);
        icons[1] = BitmapFactory.decodeResource(res,R.drawable.ic_1);
        icons[2] = BitmapFactory.decodeResource(res,R.drawable.ic_2);
        icons[3] = BitmapFactory.decodeResource(res,R.drawable.ic_3);
        iconWidth = icons[0].getWidth();
    }

    @Override
    protected void initPositionManager(){
        super.initPositionManager();
        mPm = new MutiPointManager(r,rx,ry,iconWidth,icons.length,START_ANGLE,END_ANGLE);
    }

    @Override
    public void surfaceCreated(SurfaceHolder holder) {
        drawByAngle(START_ANGLE,icon);
    }
```

我们先从触摸事件开始讲起：

```
@Override
public boolean onTouch(View v,MotionEvent event) {
    if(mIsBlocked){
        return false;
    }
    float x = event.getX();
    float y = event.getY();
    switch(event.getAction()){
```

```
        case MotionEvent.ACTION_DOWN:
            icon = mPm.isTouchedByAbsolutePoint(new Point(x,y));
            if(icon >= 0){
                isToDraw = true;
            }
            break;
        case MotionEvent.ACTION_MOVE:
            if(isToDraw){
                draw(x,y,icon);
            }
            break;
        case MotionEvent.ACTION_UP:
            if(isToDraw){
                float zeroAngle = mPm.getZeroAngleByAbsolutePoint(
                        new Point(x,y),icon);
                startThread(zeroAngle);
            }
            break;
        }
        return true;
    }
```

这里的对象 mPm 为继承自 SinglePointManager 的类：

```
public class MutiPointManager extends SinglePointManager{

    //图标个数
    private int iconCount;
    //圆周被平分角度
    private float pAngle;
    //每个图标的中心在相对坐标系下的坐标
    private Point[] relativeIconPoint;

    public MutiPointManager(float r,float rx,float ry,float iconWidth,
                int iconCount,float startAngle,float endAngle)
    {
        super(r,rx,ry,iconWidth);

        this.iconCount = iconCount;
        //总角度
        float angle = endAngle - startAngle;
        if(Math.abs(angle - 2 * Math.PI) <= 0.1 || Math.abs(angle + 2 * Math.PI) <= 0.1){
            this.pAngle = (float)angle/(iconCount);
        }
        //当总角度 = 2PI 时,由于第一个图标和最后一个图标重叠了,故图标的个数减一
        else{
            this.pAngle = (float)angle/(iconCount - 1);
        }
        //初始化各个图标的中心在相对坐标系下的坐标
        relativeIconPoint = new Point[iconCount];
```

```
        saveRelativeIconPoint(startAngle);
    }

    /**
     * 保存各个图标的中心在相对坐标系下的坐标
     * @param angle 在相对坐标上与 y 轴正半轴的夹角
     */
    public void saveRelativeIconPoint(float angle){
        for(int i = 0; i < iconCount; i++){
            float x = r * FloatMath.sin(angle + pAngle * i);
            float y = r * FloatMath.cos(angle + pAngle * i);
            Point p = new Point(x,y);
            relativeIconPoint[i] = p;
        }
    }
```

pAngle 的含义如图 5-9 所示；图中的四个图标的中心表示 relativeIconPoint。

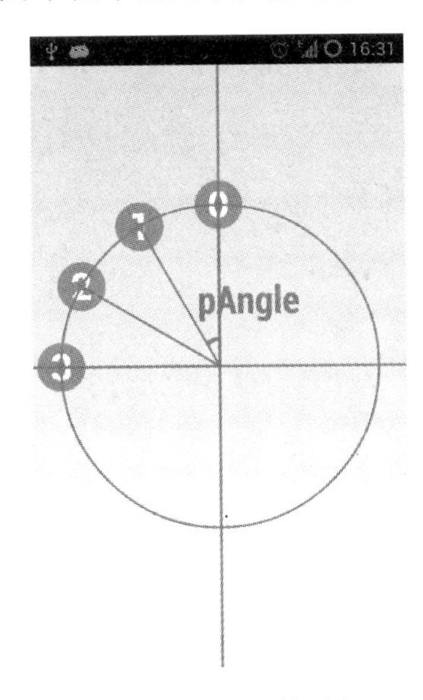

图 5-9　pAngle 的示意

首先来看 ACTION_DOWN 的事件：

```
icon = mPm.isTouchedByAbsolutePoint(new Point(x,y));
if(icon >= 0){
    isToDraw = true;
}
```

icon 表示每次拖动过程中被触摸的图标的序号，isToDraw 的含义和之前章节的一样，用来控制是否绘图：

```
protected int icon;
protected boolean isToDraw = false;
```

函数 isTouchedByAbsolutePoint 的定义如下：

```
/**
 * @param p 绝对坐标下的触摸坐标
 * @return 被触摸的图标 index, -1 表示未触摸到任何图标
 */
public int isTouchedByAbsolutePoint(Point p)
{
    Point relativePoint = getRelativePointByAbsolutePoint(p);
    float x = relativePoint.getX();
    float y = relativePoint.getY();
    for(int i = 0; i < relativeIconPoint.length; i++)
    {
        Point iconPoint = this.getIconPoint(relativeIconPoint[i],false);
        if(x > iconPoint.getX() && x < iconPoint.getX() + iconWidth &&
                y < iconPoint.getY() && y > iconPoint.getY() - iconWidth)
        {
            return i;
        }
    }
    return -1;
}
```

事件 ACTION_MOVE 的代码就几行：

```
if(isToDraw){
    draw(x,y,icon);
}
```

在绘图的过程中，需要对坐标进行转换、获取。与单个图形的坐标计算类似，也需要进行多次的坐标转换。不同的是，在这个例子中，我们需要获取多个绘图坐标，然后进行绘图。

```
protected void draw(float x,float y,int ic){
    Point p = new Point(x,y);
    Point[] pts = mPm.getResultPointsByAbsolutePoint(p,ic);
    float angle = mPm.getZeroAngleByAbsolutePoint(p,ic);
    //将 PI 转化为 180
    angle = getAngle(angle);

    Canvas canvas = mHolder.lockCanvas();
    canvas.drawColor(Color.TRANSPARENT,Mode.CLEAR);

    for(int i = 0; i < pts.length; i++){
        Point pt = pts[i];

        matrix.reset();
        matrix.setTranslate(pt.getX(),pt.getY());
        matrix.preRotate(angle,
              (float)icons[i].getWidth()/2,(float)icons[i].getWidth()/2);
```

```
            canvas.drawBitmap(icons[i],matrix,paint);

        }

        /*
         * 绘制辅助图形
         * 可以省去
         */
        drawUtils(canvas,x,y,ic);

        mHolder.unlockCanvasAndPost(canvas);
    }
```

获取绘图坐标与单个获取绘图坐标的原理类似,但是有两个不同的地方:

(1) 获取第0号图标在相对坐标下与y轴正半轴的夹角:

```
/**
 * @param pt 相对坐标
 * @param index 图标 index
 * @return 0 号图标在相对坐标系下与 y 轴正半轴的夹角
 */
protected float getZeroAngleByRelativePoint(Point pt,int index){
    float angle = getAngleByRelativePoint(pt);
     //pAngle 每个图标之间的夹角
    angle -= index * pAngle;
    return angle;
}
```

(2) 根据角度计算坐标时,第i个图标的角度为:

```
//pAngle 每个图标之间的夹角
float iconAngle = angle + pAngle * i;
```

获取绘图坐标:

```
/**
 * @param pt 绝对坐标下的触摸坐标
 * @param index 触摸的图标 index
 * @return 绘图坐标
 */
public Point[] getResultPointsByAbsolutePoint(Point pt,int index){

    Point[] ret = new Point[iconCount];
    //将绝对坐标转换成相对坐标
    Point relativePoint = getRelativePointByAbsolutePoint(pt);
    //获取第 0 号图标在相对坐标下与 y 轴正半轴的夹角
    float angle = getZeroAngleByRelativePoint(relativePoint,index);
    //当角度为 2PI 时,取零
```

```
    if(Math.abs(angle - 2 * Math.PI) <= 0.01){
        angle = 0;
    }

    for(int i = 0; i < iconCount; i++){
        float iconAngle = angle + pAngle * i;
        //计算相对坐标系下的坐标
        relativePoint = new Point(
              FloatMath.sin(iconAngle) * r,FloatMath.cos(iconAngle) * r);
        //获取绝对坐标系下的坐标
        Point absolutePoint = getAbsolutePointByRelativePoint(relativePoint);
        //获取图标的绘图坐标(左上角)
        Point resultPoint = getIconPoint(absolutePoint,true);
        ret[i] = resultPoint;
    }
    return ret;
}
```

ACTION_UP 时：

```
if(isToDraw){
    float zeroAngle = mPm.getZeroAngleByAbsolutePoint(
          new Point(x,y),icon);
    startThread(zeroAngle);
}
protected void startThread(float zeroAngle){
    new RollbackThread(
        zeroAngle,
        0.1f,
        START_ANGLE)
    .start();
}
protected class RollbackThread extends Thread{

    protected float destination = 0;
    protected int direction = 1;
    protected boolean isRun = true;
    protected float zeroAngle;
    protected float movingOffset;

    /**
     * @param angle 第 0 号图标开始时的旋转角度
     * @param movingOffset 每次旋转的角度
     * @param destination 最终的旋转角度
     */
    public RollbackThread(float angle,float movingOffset,float destination){
        if(angle < 0){
            angle += Math.PI * 2;
        }else if(angle > Math.PI * 2){
            angle -= Math.PI * 2;
```

```
        }
        this.zeroAngle = angle;
        this.movingOffset = movingOffset;
        this.destination = destination;
        direction = getDirection(angle,destination);
    }

    public void run(){
        mIsBlocked = true;
        while(isRun){
            if(Math.abs(zeroAngle - destination) <= movingOffset ||
                    Math.abs(zeroAngle - destination) >= Math.PI * 2){
                isRun = false;
                drawByAngle(destination,0);
                mPm.saveRelativeIconPoint(destination);
            }else{
                zeroAngle += direction * movingOffset;
                drawByAngle(zeroAngle,0);
            }
        }
        reset();
    }

    /**
     * @param angle
     * @param destination
     * @return 旋转的方向
     */
    private int getDirection(float angle,float destination){
        float a = angle - destination;
        return FloatMath.sin(a) > 0? -1:1;
    }
}

/**
 * 重置所有的控制
 */
protected void reset(){
    icon = -1;
    isToDraw = false;
    mIsBlocked = false;
}
```

图标移回后自动旋转

最后，我们添加一点小动画，在图标回到原点时旋转 90°，回复到原来的角度。

```
public class SpinBackView extends ImageSetView{

    public SpinBackView(Context context,AttributeSet attrs) {
```

```
        super(context,attrs);
    }

    @Override
    protected void startThread(float zeroAngle){
        new ImageSetRollbackThread(
                zeroAngle,
                0.1f,
                START_ANGLE)
        .start();
    }

    private class ImageSetRollbackThread extends RollbackThread{

        public ImageSetRollbackThread(float angle,float movingOffset,
                float destination) {
            super(angle,movingOffset,destination);
            //TODO Auto-generated constructor stub
        }

        public void run(){
            super.run();
            //在 RollbackThread 结束之后开始图标自动旋转的线程
            new SpinThread(destination,0.1f).start();
        }
    }

    class SpinThread extends Thread{

        private boolean isRun = true;
        //最后的角度
        private float angle;
        //旋转的角度
        private float spinAngle;
        //结束的角度
        private float destination;
        private float movingOffset;

        public SpinThread(float angle,float movingOffset){
            this.angle = angle;
            this.spinAngle = angle;
            this.movingOffset = movingOffset;
            destination = (float) (angle + Math.PI * 2);
        }

        public void run(){
            mIsBlocked = true;
            while(isRun){
                if(Math.abs(spinAngle - destination) <= movingOffset){
                    isRun = false;
                    drawSpinByAngle(angle,0,destination);
                }else{
```

```
                    spinAngle + = movingOffset;
                    drawSpinByAngle(angle,0,spinAngle);
                }
            }
            reset();
        }
    }
    private void drawSpinByAngle(float angle,int ic,float spinAngle){

        Point pt = mPm.getAbsolutePointByAngle(angle);
        draw(pt.getX(),pt.getY(),ic,getAngle(spinAngle));

    }

    /**
     * 绘制图形
     * @param x
     * @param y
     * @param ic
     * @param spinAngle 图形旋转的角度
     */
    protected void draw(float x,float y,int ic,float spinAngle){
        Point p = new Point(x,y);
        Point[] pts = mPm.getResultPointsByAbsolutePoint(p,ic);

        Canvas canvas = mHolder.lockCanvas();
        canvas.drawColor(Color.TRANSPARENT,Mode.CLEAR);

        for(int i = 0; i < pts.length; i++){
            Point pt = pts[i];

            matrix.reset();
            matrix.setTranslate(pt.getX(),pt.getY());
            //图标旋转
            matrix.preRotate(spinAngle,
                    (float)icons[i].getWidth()/2,(float)icons[i].getWidth()/2);
            canvas.drawBitmap(icons[i],matrix,paint);
        }
        /*
         * 绘制辅助图形
         */
        drawUtils(canvas,x,y,ic);
        mHolder.unlockCanvasAndPost(canvas);
    }
}
```

第五节　时钟控件的实现

还记不记得我们在第 3 章提到的时钟控件？这个时钟使用了本节的知识，因此我把它移到这一节中来讲解。

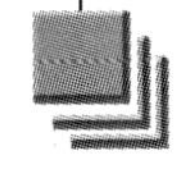

要实现这样一个时钟，需要这么几个图片资源，如图 5-10 所示。

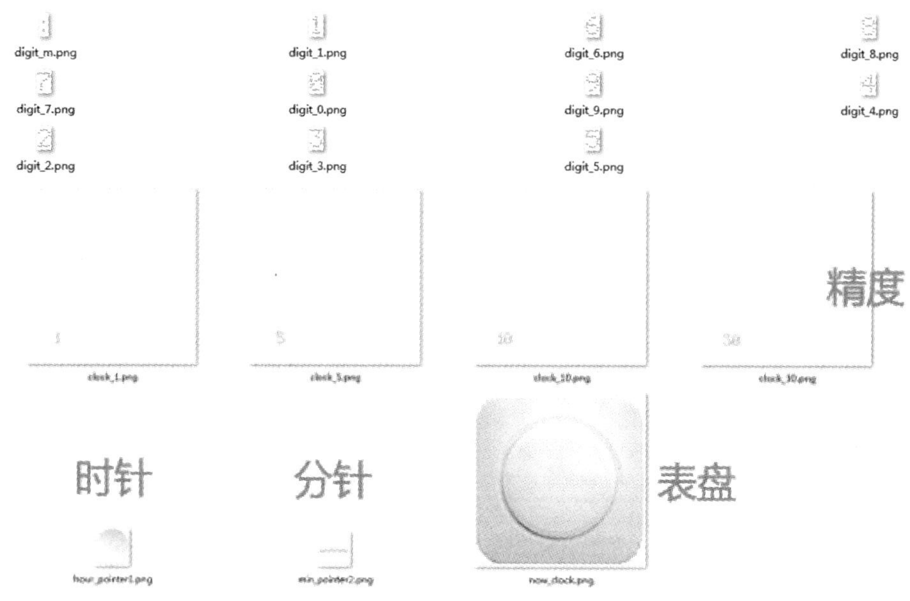

图 5-10　图片资源

为了实现这样一个美观(奇葩)的时钟，需要解决三个大问题：

(1) 根据时间，计算出时针、分针的角度。

(2) 根据时针的角度计算出时间。

(3) 设置精度。

本节就以这几个问题出发讲解时钟的实现。

根据时间绘制时针分针

这个问题的表述为：给定任意一个时间(24 小时制)h:m，计算时针与分针的角度。为了简化问题，我们先来看分针的计算。

分针的角度只跟分也就是 m 的值有关。一小时有 60 分钟，一圈有 360°，那么每一分钟分针就走过了 360°/60=6°。因此有静态变量：

```
float ANGLE_PER_MIN = (float) (Math.PI * 2/ROUND_MIN);
```

同理，每一小时时针走过 360°/12=30°：

```
float ANGLE_PER_HOUR = (float) (Math.PI * 2/ROUND_HOUR);
```

因此获取分针角度的函数也很简单：

```
private float getAngleOfMin(int min){
    return (float) (min * ANGLE_PER_MIN);
}
```

由于分针需要旋转，因此绘制分针的函数多了一个 matrix 的操作：

```
private void drawMin(TimeVo time,Canvas canvas){
    //获取角度
    float angleOfMin = getAngleOfMin(time.getMin());
    //获取绘图坐标
    Point minPoint = minPm.getResultPointByAngle(angleOfMin);
    drawMin(minPoint,angleOfMin,canvas);
}
private void drawMin(Point minPoint,float angle,Canvas canvas){
    //π->180
    angle = (float) ((angle * (ROUND_ANGLE/2))/Math.PI);
    //旋转
    matrix.reset();
    matrix.setTranslate(minPoint.getX(),minPoint.getY());
    matrix.preRotate(angle,(float)minPointerBm.getWidth()/2,
                    (float)minPointerBm.getHeight()/2);
    canvas.drawBitmap(minPointerBm,matrix,minPaint);
}
```

此外，minPm 为分针的坐标控制的对象(接下来的 hourPm 也相同)：

```
private PointManager hourPm;
private PointManager minPm;
```

PointManager 继承自 SinglePointManager：

```
public class PointManager extends SinglePointManager{

    //图标的绝对坐标
    private Point position;

    public PointManager(float r,float rx,float ry,float iconWidth) {
        super(r,rx,ry,iconWidth);
    }

    /**
     * @param absolutePoint 绝对坐标
     * @param offset 偏移
     * @return
     */
    public boolean isTouched(Point absolutePoint,int offset)
    {
        if(absolutePoint.getX() > position.getX() - offset/2 &&
                absolutePoint.getX() < position.getX() + iconWidth + offset/2 &&
                absolutePoint.getY() > position.getY() - offset/2 &&
                absolutePoint.getY() < position.getY() + (iconWidth + offset/2)){
            return true;
        }else{
```

```
            return false;
        }
    }

    @Override
    public Point getResultPointByAngle(float angle){
        //保存图标的坐标
        return position = super.getResultPointByAngle(angle);
    }
}
```

以及变量的初始化：

```
private void createPointBm(Resources res){
    //表盘
    clockBm = BitmapFactory.decodeResource(res,R.drawable.now_clock);
    clockWidth = clockBm.getWidth();
    clockHeight = clockBm.getHeight();
    clockRect = new Rect(0,0,clockWidth,clockHeight);

    //时针分针
    hourPointerBm = BitmapFactory.decodeResource(
                res,R.drawable.hour_pointer1);
    minPointerBm = BitmapFactory.decodeResource(
                res,R.drawable.min_pointer2);
    /*
     * 参数1       半径
     * 参数2       原点的x轴坐标
     * 参数3       原点的y轴坐标
     * 参数4       图标宽度
     */
    hourPm = new PointManager(HOUR_POINTER_LENGTH,((float)clockWidth)/2,
            ((float)clockHeight)/2,hourPointerBm.getWidth());
    minPm = new PointManager(MIN_POINTER_LENGTH,((float)clockWidth)/2,
            ((float)clockHeight)/2,minPointerBm.getWidth());

    //分针抗锯齿
    minPaint = new Paint();
    minPaint.setAntiAlias(true);
}
```

接下来绘制时针。时针的绘制相对于分针在计算上比较复杂，但是由于不需要旋转，所以在绘图上比较简单。首先获取时针的角度，再加上分钟带来的角度偏移：

```
private float getAngleOfHour(TimeVo time){
    /*
     * 去24小时化
     * ROUND_HOUR = 12
     */
```

```
        float angleOfHour = (float) ((time.getHour() % ROUND_HOUR) * ANGLE_PER_HOUR);
        //加上分钟带来的角度偏移
        angleOfHour + = time.getMin() * (ANGLE_PER_HOUR/ROUND_MIN);
        return angleOfHour;
    }
```

绘图：

```
    private void drawHour(TimeVo time,Canvas canvas){
        float angleOfHour = getAngleOfHour(time);
        //获取绘图坐标
        Point hourPoint = hourPm.getResultPointByAngle(angleOfHour);
        drawHour(hourPoint,canvas);
    }
    private void drawHour(Point hourPoint,Canvas canvas){
        canvas.drawBitmap(hourPointerBm,hourPoint.getX(),hourPoint.getY(),null);
    }
```

根据时针位置绘制分针

根据时针绘制分针的关键是根据时针的角度获取分针的角度。假设时间为 1:20 分，那么此时的时针角度应该是 1×30＋(20÷60)×30，也就是 30×(1＋20°÷60)＝40；那么对这个公式进行逆运算就是：

(1) 40 除以 30 取余得 10。

(2) 10 除以 30 得 1/3。

(3) 1/3 乘以 360°得 120°。

因而有：

```
    private float getMinAngleByHour(double angleOfHour){
        float angleLeft = (float) (angleOfHour % ANGLE_PER_HOUR);
        float angleOfMin = angleLeft/ANGLE_PER_HOUR;
        return (float) (angleOfMin * Math.PI * 2);
    }
```

再接下来，由于有精度的存在，我们需要获取精度值，关于精度值在这里我就不多说了，看注释中的几个例子：

```
    /**
     * 获取精度值
     * 例如,getNearestValue(47,5) = 45; getNearestValue(47,10) = 40;
     * getNearestValue(0.47,0.05) = 0.45
     * @param value
     * @param by
     * @return
     */
```

```
private double getNearestValue(double value,double by){
    double divide = by;
    value = value/divide;
    by = by/divide;
    double result = (int)value/(int)by * by;
    return result * divide;
}
```

因而有绘制时针分针的代码如下：

```
//根据坐标获取相对坐标的角度
double angleOfHour = hourPm.getAngleByAbsolutePoint(p);
//获取精度值
angleOfHour = getNearestValue(angleOfHour,hourGapAngle);
Point hourPoint = hourPm.getResultPointByAngle((float) angleOfHour);
drawHour(hourPoint,canvas);

float angleOfMin = getMinAngleByHour(angleOfHour);
//获取精度值
angleOfMin = (float) getNearestValue(angleOfMin,minGapAngle);
Point minPoint = minPm.getResultPointByAngle((float) angleOfMin);
drawMin(minPoint,angleOfMin,canvas);
```

再接下来就是0点问题了，也就是当时针指向12点时，是12点还是0点。

```
previousMin = angleOfMin;
/**
 * 越过0点,
 * 11点之后是12点
 * 23点之后是0点
 */
if(Math.abs(previousHour - angleOfHour) > Math.PI * 3/2){
    adding = (adding + 12) % 24;
}
previousHour = angleOfHour;

//设置当前时间
setTimeByAngle(angleOfMin,angleOfHour);
```

保存当前时间：

```
private void setTimeByAngle(double angleOfMin,double angleOfHour){
    int hour = (int) (angleOfHour/ANGLE_PER_HOUR);
    int min = (int) (angleOfMin/ANGLE_PER_MIN);
    hour + = adding;
    currentTime.setHour(hour);
    currentTime.setMin(min);
}
```

在绘图的时候，根据这个保存的时间 currentTime 来绘制数字时间：

```
private void drawDigitalTime(Canvas canvas,Paint pt){
    int hour = currentTime.getHour();
    int min = currentTime.getMin();
    int hour1 = hour/10;
    int hour2 = hour % 10;
    int min1 = min/10;
    int min2 = min % 10;
    int[] indexs = new int[]{hour1,hour2,10,min1,min2};   //hh:mm
    for(int i = 0; i < indexs.length; i++){
        canvas.drawBitmap(digitBm[indexs[i]],positioDdigitX[i] + DIGIT_OFFSET,
                DIGIT_OFFSET,pt);
    }
}
```

还有最后一个小问题，音效播放的控制：

```
//是否播放音效
if(Math.abs(previousMin - angleOfMin) <= 0.01f){
    isToPlaySoundEffect = false;
}
previousMin = angleOfMin;
```

第六节　扩展学习——浮窗应用

WindowManager

在 Android 系统中，所有的窗口(包括 Activity)都是由 WindowManager 来控制的。我们在使用 Activity 的时候，系统帮我们调用了 WindowManager 来控制 Activity 的创建，因而虽然我们不需要使用它，但它是客观存在的。因此，要建立一个浮窗，可以通过 WindowManager 来添加一个 View，从而绕开 Activity，实现效果。

在 Activity 中，我们通过 getSystemService 的方法来获取 WindowManager 对象：

```
WindowManager wm = (WindowManager) getSystemService(WINDOW_SERVICE);
```

WindowManager 对象是一个全局变量，在 Android 系统中只有一个，因此我们不管在哪里获取这个变量都是 equal 的。

WindowManager 的重要方法有：

- addView (View view,ViewGroup.LayoutParams params)；
- removeView (View view)；
- updateViewLayout (View view,ViewGroup.LayoutParams params)。

在这里需要一个 LayoutParams 的对象，这个对象提供了视图的参数，主要的参数有：

- WindowManager. LayoutParams. type；
- WindowManager. LayoutParams. flag；
- WindowManager. LayoutParams. gravity；
- WindowManager. LayoutParams. x；
- WindowManager. LayoutParams. y。

浮窗实例

先来看一个最简单的例子，不用 Layout，直接在 Java 文件上写代码：

```
public class MainActivity extends Activity {

    @Override
    protected void onCreate(Bundle savedInstanceState) {
        super.onCreate(savedInstanceState);

        WindowManager wm = (WindowManager) getSystemService(WINDOW_SERVICE);
        ImageView view = new ImageView(this,null);
        view.setImageResource(R.drawable.ic_launcher);
        /*
         * 注意:
         * 这个 LayoutParams 是  android.view.WindowManager.LayoutParams
         */
        LayoutParams wmParams = new WindowManager.LayoutParams();
        /**
         * TYPE_SYSTEM_ERROR 最顶层,可以移动
         */
            wmParams.type = LayoutParams.TYPE_SYSTEM_ERROR;
            /*
             * 如果没有这个 flag,整个 View 会覆盖全屏幕
             * 不信你试试
             */
            wmParams.flags = LayoutParams.FLAG_NOT_FOCUSABLE;
            //设置重心
            wmParams.gravity = Gravity.LEFT|Gravity.TOP;
            wmParams.x = 0;
            wmParams.y = 0;
            wmParams.width = icon.getWidth();
            wmParams.height = icon.getHeight();

        wm.addView(view,wmParams);
    }

}
```

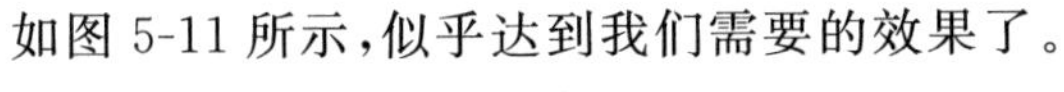
如图 5-11 所示，似乎达到我们需要的效果了。

图 5-11　浮窗

按返回键退出，whoops，报错了：

```
Activity com.example.demo_floatview.MainActivity has leaked window android.widget.ImageView
{4186ba38 V.ED.... ........I.0,0 - 72,72}that was originally added here
```

按照字面的理解是，ImageView 发生泄漏了。我们在使用 BroadcastReceiver 只注册不注销的时候也会出现泄漏的错误。有了这个经验，我们可以在 Activity 销毁的时候加上：

```
@Override
public void onDestroy(){
    super.onDestroy();
    wm.removeView(view);
}
```

再运行的时候，就不会报错了。

但是仔细想想好像有什么不对。如果程序退出，视图就销毁，那么浮窗又有什么意义？

没事，我们还有办法，那就是把 addView 的方法放在需要添加的视图里面，而不是在 Activity 中。这时，我们就需要重写一个 SurfaceView（顺便把难看的黑色背景去掉）。

因为浮窗程序中只能允许一个浮窗的存在，因此我们使用单例模式。Activity 在调用时不直接调用构造函数，而是通过静态方法来实例化这个对象：

```
public class FloatView extends SurfaceView implements SurfaceHolder.Callback{

    //使用单例模式，只允许系统中存在一个浮窗
    protected static FloatView thisView;

    public static FloatView createView(Context context,AttributeSet attrs){
```

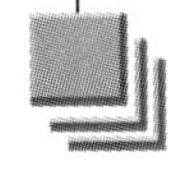

```
        if( thisView == null){
            thisView = new FloatView(context,attrs);
        }
        return thisView;
    }
```

构造函数对外部不可见：

```
protected FloatView(Context context,AttributeSet attrs) {
    super(context,attrs);
    icon = BitmapFactory.decodeResource(
                context.getResources(),R.drawable.ic_launcher);
    wm = (WindowManager)context.getApplicationContext().
                getSystemService(Context.WINDOW_SERVICE);
    iniWmParams();

    sfh = this.getHolder();
    sfh.addCallback(this);

    //透明
    setZOrderOnTop(true);
    sfh.setFormat(PixelFormat.TRANSLUCENT);
}
private void iniWmParams()
{
    wmParams = new WindowManager.LayoutParams();
    /**
     * TYPE_SYSTEM_ERROR 最顶层,可以移动
     */
        wmParams.type = LayoutParams.TYPE_SYSTEM_ERROR;
        //如果没有这个 flag,整个 View 会覆盖系统界面
     wmParams.flags = LayoutParams.FLAG_NOT_FOCUSABLE;
     wmParams.gravity = Gravity.LEFT|Gravity.TOP;
     wmParams.x = 0;
     wmParams.y = 0;

     wmParams.width = icon.getWidth();
     wmParams.height = icon.getHeight();
}
```

绘图的初始化：

```
@Override
public void surfaceCreated(SurfaceHolder sfh) {
    Canvas canvas = sfh.lockCanvas();
    canvas.drawColor(Color.TRANSPARENT,Mode.CLEAR);
    canvas.drawBitmap(icon,0,0,new Paint());
    sfh.unlockCanvasAndPost(canvas);
}
```

最后提供方法供 Activity 调用：

```
public void showView(){
    wm.addView(thisView,wmParams);
}
public void destroyView(){
    wm.removeView(thisView);
}
```

Activity 的代码非常简单：

```
public class FloatActivity extends Activity {
    @Override
    protected void onCreate(Bundle savedInstanceState) {
        super.onCreate(savedInstanceState);
        FloatView.createView(this,null).showView();
    }
}
```

现在，这个浮窗就自由地浮动在手机上方，再也不用担心侧漏了。

拖动浮窗的实现

在很多情况下，浮窗是可以响应触摸事件并被拖动的，我们可以通过设置监听器来实现这个效果。值得注意的是，在获取坐标时，是使用 getRawX()方法而非 getX()，这是因为 WindowManager 在管理 View 的位置时，是以整个屏幕为视图的。重设完坐标值之后调用 updateViewLayout 方法：

```
public class MovingView extends FloatView implements View.OnTouchListener {

    protected MovingView(Context context,AttributeSet attrs) {
        super(context,attrs);
        setOnTouchListener(this);
    }

    /*
     *注意
     *这里不能用@Override
     *因为它是静态方法
     */
    public static MovingView createView(Context context,AttributeSet attrs){
        if(thisView == null){
            thisView = new MovingView(context,attrs);
        }
        return (MovingView) thisView;
    }
    @Override
    public boolean onTouch(View arg0,MotionEvent event) {
        if(event.getAction() == MotionEvent.ACTION_MOVE){
            /*
```

```
                *更改布局的位置
                *相对于整个界面
                *所以是 getRawX()跟 getRawY
                */
               int x = (int) event.getRawX();
               int y = (int) event.getRawY();
              wmParams.x = x;
              wmParams.y = y;
              wm.updateViewLayout(thisView,wmParams);
          }
          return true;
      }
  }
```

第六章 Easearch

类 Launch 的搜索应用

清单

Demo 代码：

```
\demo\Demo_Easearch
\demo\Demo_Easearch_Intent
```

实例代码：

```
\source_codes\Earch
```

Target SDK：

```
Android 2.3.3
```

第一节 产品介绍

这个项目始于 Terminal 项目，如图 6-1 所示。当时的想法是做一个基于代码行的减弱用户体验的类 Launch 应用，打电话、发短信、打开应用等等所有的操作都由代码执行。

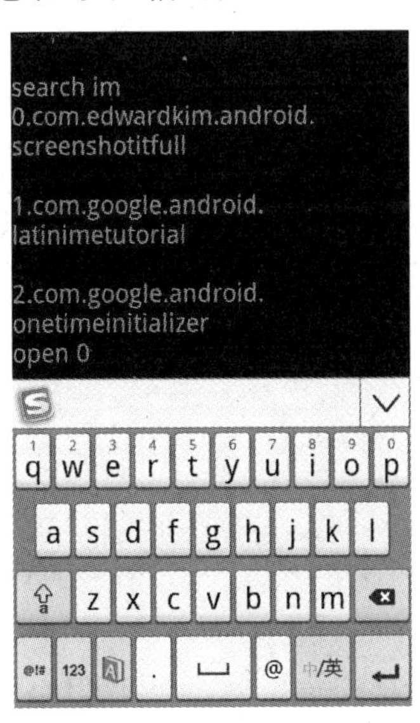

图 6-1 Terminal 项目

这个项目后来演化为由智能控制应用，如图 6-2 所示，最终还是胎死腹中。

后来想一想，这种减弱用户体验的应用实在是太逆天了，应用的终极目标就是让手机好用。因而它最终演化成了现在的 Easearch，如图 6-3 所示。Easearch 是 easy 跟 search 两个单词的组合。它的概念是手机的操作都基于搜索，搜索联系人、应用、网页以及单词等等。除了搜索之外，还提供了 WiFi、数据流量等快捷开关。

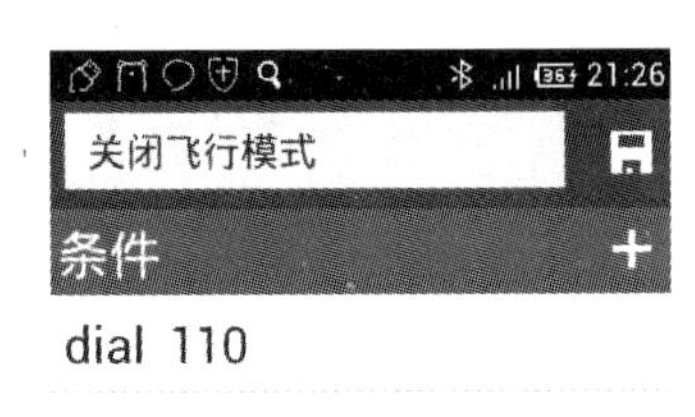

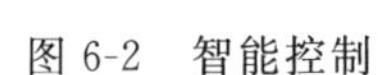
图 6-2　智能控制

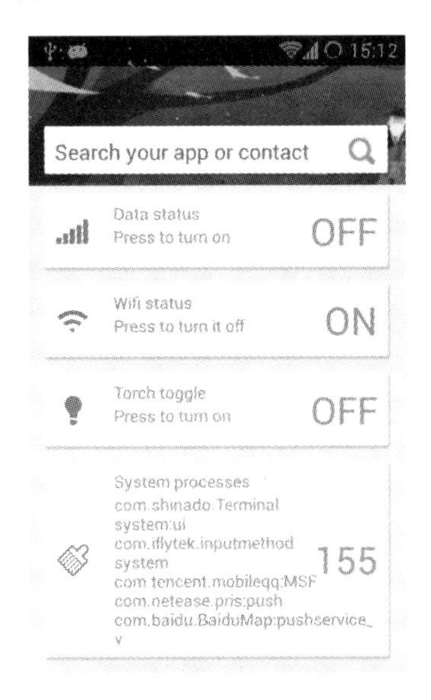

图 6-3　Easearch 应用

需求分析

Easearch 有以下几个功能：

- 搜索联系人；
- 滑动联系人列表，进入联系人短信界面；
- 滑动联系人列表，复制联系人信息；
- 搜索应用；
- 滑动应用列表，进入应用信息界面；
- 滑动应用列表，删除应用；
- Wifi、数据流量、蓝牙、手电筒开关；
- 一键清理内存；
- 刷新。

界面设计

Easearch 模仿了 Google Now 卡片式的界面风格，如图 6-4 所示。

搜索结果栏为有一个动态的展开效果：当没有输入或者没有相关的信息时，该栏不显示；当出现相关信息时，会有展开的动态效果，如图 6-5 所示。

在搜索结果栏中，用户可以左右滑动，进入不同功能，如图 6-6 所示。

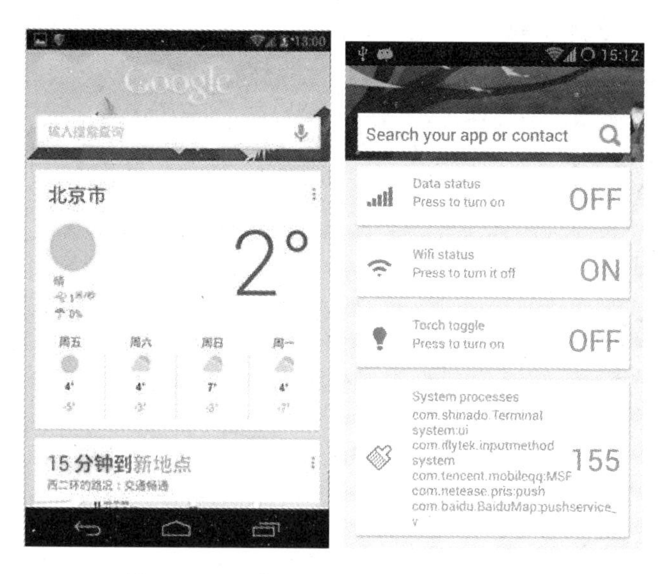

图 6-4　Google Now(左)和 Easearch(右)

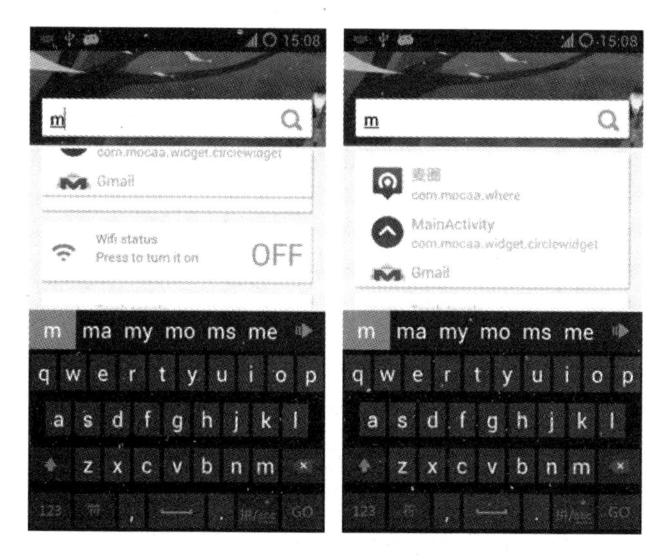

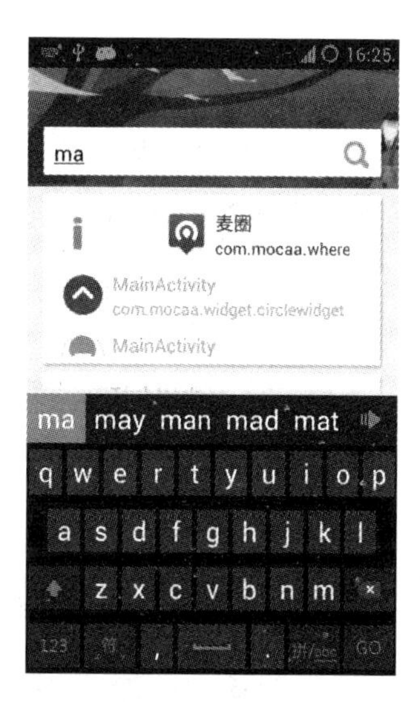

图 6-5　列表展开　　　　图 6-6　左右滑动列表

第二节　调用系统界面/服务

Android 提供了使用 Intent 来调用系统界面的 API，这些 API 都是通过 Intent 跟 URI 的搭配来实现的。本节将会介绍 Android 中调用系统界面的实现。

隐式 Intent

我们在第 2 章的时候介绍了 Intent，现在我们来继续介绍显式 Intent 和隐式 Intent 的区别。

在第 2 章中，如果要从一个 Activity 跳到另一个 Activity，可以这样：

```
Intent intent = new Intent();
intent.setClass(this,NewActivity.class);
startActivity(intent);
```

这就是显式 Intent，明确指明了要跳转的 Activity。而隐式 Intent 则是通过 action、data、category 等条件来跳转 Activity，如：

```
Intent intent = new Intent();
intent.setAction(Intent.ACTION_DIAL);
Uri uri = Uri.parse("tel:13888888888");
intent.setData(uri);
/*
 * 相当于 Intent intent = new Intent(Intent.ACTION_DIAL,uri);
 * 第一个参数 action,第二个参数 data
 */
startActivity(intent);
```

实际上，隐式 Intent 属于跨进程通信的一种方式，也就是说可以从应用 A 调用应用 B 的 Activity，而显式 Intent 只能调用到本应用的 Activity。这也就是 Android 中调用系统界面的原理。

我们来做个小 demo，通过指定 action 的方式来打开另一个应用的 Activity。

在应用 A 中：

```
Intent intent = new Intent();
intent.setAction("com.shinado.demo.easearch.intent");
startActivity(intent);
```

在应用 B 的 AndroidManifest 中对需要启动的 Activity 增加一个 intent-filter：

```
<intent-filter>
    <action android:name="com.shinado.demo.easearch.intent"/>
    <category android:name="android.intent.category.DEFAULT"/>
</intent-filter>
```

这里的 action 与应用 A 中的 action 是一致的。而 android.intent.category.DEFAULT 这个 category 是为了保证隐式 Intent 可以 match 到这个 filter，如果去掉这个 category 就会报错：

```
Caused by:android.content.ActivityNotFoundException:No Activity found to handle Intent{act=
com.shinado.demo.easearch.intent}
```

使用隐式 Intent 调用系统界面

笔者在这里列出常用的调用系统界面的代码，方便读者查阅。

调用拨号界面：

```
Uri uri = Uri.parse("tel:13888888888");
Intent intent = new Intent(Intent.ACTION_DIAL,uri);
startActivity(intent);
```

直接拨号：

```
Uri uri = Uri.parse("tel:13888888888");
Intent intent = new Intent(Intent.ACTION_CALL,uri);
startActivity(intent);
```

常见异常

直接拨号需要权限 android. permission. CALL_PHONE，否则会抛出异常

```
Caused by:java.lang.SecurityException:Permission Denial:starting Intent{act = android.intent.action.CALL dat = tel:xxxxxxxxxx cmp = com.android.phone/.OutgoingCallBroadcaster} fron ProcessRecord{41ddfc38 25527:com.example.demo_eas earch/u0a10104()pid = 25527.uid = 10104)reouires android.dormission.CALL PHONE
```

调用通话记录界面：

```
Intent intent = new Intent();
intent.setAction(Intent.ACTION_CALL_BUTTON);
startActivity(intent);
```

进入联系人界面：

```
Intent intent = new Intent();
intent.setAction(Intent.ACTION_PICK);
intent.setData(Phone.CONTENT_URI);
startActivity(intent);
```

编辑第 5 号联系人：

```
Uri uri = Uri.parse("content://com.android.contacts/contacts/" + "5");
Intent intent = new Intent(Intent.ACTION_EDIT,uri);
startActivity(intent);
```

添加联系人：

```
Intent intent = new Intent(Intent.ACTION_INSERT_OR_EDIT);
intent.setType("vnd.android.cursor.item/contact");
intent.putExtra(android.provider.ContactsContract.Intents.Insert.NAME,"name");
intent.putExtra(android.provider.ContactsContract.Intents.Insert.COMPANY,"c");
intent.putExtra(android.provider.ContactsContract.Intents.Insert.PHONE,"num");
intent.putExtra(android.provider.ContactsContract.Intents.Insert.PHONE_TYPE,1);
startActivity(intent);
```

比较诡异的是，调用了这段代码之后，会跳到联系人的页面，单击“创建新联系人”才会有我们预置的信息，如图 6-7 所示。

图 6-7　联系人页面

进入发短信界面：

```
//需要发短信的号码
Uri uri = Uri.parse("smsto:13888888888");
Intent intent = new Intent(Intent.ACTION_SENDTO,uri);
//短信内容
intent.putExtra("sms_body","短信内容");
startActivity(intent);
```

直接发送短信(慎用)：

```
SmsManager smsManager = SmsManager.getDefault();
List<String> divideContents = smsManager.divideMessage("hello");
for (String text : divideContents) {
    smsManager.sendTextMessage("13888888888",null,text,null,null);
}
```

常见异常

发送短信需要权限 android.permission.SEND_SMS，否则会抛出异常

```
Caused by: java.lang.SecurityException: Sending SMS messate: uid 10104 does not have
android.permission.SEND SMS.
```

访问网页：

```
Uri uri = Uri.parse("http://www.google.com");
Intent it  = new Intent(Intent.ACTION_VIEW,uri);
startActivity(it);
```

查看位置：

```
Uri uri = Uri.parse("geo:28.172,112.93");
Intent it = new Intent(Intent.ACTION_VIEW,uri);
startActivity(it);
```

路径规划：

```
Uri uri = Uri.parse("https://maps.google.com/maps?f = d" +
                  "&saddr = 28.168,112.93" +
                  "&daddr = 28.172,112.926");
Intent it = new Intent(Intent.ACTION_VIEW,uri);
startActivity(it);
```

删除应用：

```
//应用的包名
Uri uri = Uri.fromParts("package","com.shinado.Terminal",null);
Intent it = new Intent(Intent.ACTION_DELETE,uri);
startActivity(it);
```

查看应用信息：

```
Intent intent = new Intent(Settings.ACTION_APPLICATION_DETAILS_SETTINGS);
//应用的包名
Uri uri = Uri.fromParts("package","com.shinado.Terminal",null);
intent.setData(uri);
startActivity(intent);
```

进入蓝牙/WiFi/GPRS 设置界面：

```
//蓝牙设置
Intent intent = new Intent(Settings.ACTION_BLUETOOTH_SETTINGS);
startActivity(intent);
//GPRS 设置
Intent intent = new Intent(Settings.ACTION_WIRELESS_SETTINGS);
startActivity(intent);
//wifi 设置
Intent intent = new Intent(Settings.ACTION_WIFI_SETTINGS);
startActivity(intent);
```

调用系统功能

开关蓝牙

```
BluetoothAdapter bluetooth = BluetoothAdapter.getDefaultAdapter();
if(bluetooth != null) {
```

```
    if(bluetooth.isEnabled()){
        bluetooth.disable();
    }else{
        bluetooth.enable();
    }
}
```

常见异常

开关蓝牙需要权限 android.permission.BLUETOOTH_ADMIN 以及 android.permission.BLUETOOTH，否则会抛出异常

```
Caused by: java.lang.SecurityException: Need BLUETOOTH ANDIN permission: Neit her user 10104 nor current process has android.permission.BLUETOOTH_ADMIN.
Caused by: javalang.SecurityException: Need BLUETOOTH permission: Neither user 10104 nor current process has android.permission.BLUETOOTH.
```

开关 GPRS：

由于开关 GPRS 的 API 被 Google 封掉了，我们只能用反射的方式来获取这个方法：

```
public void toogleGprs()
{
    ConnectivityManager mCM =
          (ConnectivityManager)getSystemService(Context.CONNECTIVITY_SERVICE);
    boolean isOpen = isDataOn(mCM);
    if(isOpen){
        setGprsEnable(mCM,false);
    }else{
        setGprsEnable(mCM,true);
    }
}
//检测 GPRS 是否打开
private boolean isDataOn(ConnectivityManager mCM)
{
    Class cmClass   = mCM.getClass();
    Class[] argClasses   = null;
    Object[] argObject   = null;
    Boolean isOpen = true;
    try{
        Method method = cmClass.getMethod("getMobileDataEnabled",argClasses);
        isOpen = (Boolean) method.invoke(mCM,argObject);
    }catch (Exception e){
        e.printStackTrace();
    }
    return isOpen;
}

//开启/关闭 GPRS
private void setGprsEnable(ConnectivityManager mCM,boolean isEnable)
{
```

```
    Class cmClass = mCM.getClass();
    Class[] argClasses = new Class[1];
    argClasses[0] = boolean.class;

    try{
        Method method = cmClass.getMethod("setMobileDataEnabled",argClasses);
        method.invoke(mCM,isEnable);
    } catch (Exception e){
        e.printStackTrace();
    }
}
```

常见异常

开关 GPRS 需要权限 android.permission.CHANGE_NETWORK_STATE，否则会抛出异常

```
Caused by: java.lang.SecurltyException: ConnectlvityService: Neither user 10104 nor current
process has android.permission.CHANGE_NETWORK_STATE.
```

开关 WiFi：

```
WifiManager wm = (WifiManager)getSystemService(Context.WIFI_SERVICE);
switch(wm.getWifiState()){
case WifiManager.WIFI_STATE_DISABLING:
    break;
case WifiManager.WIFI_STATE_DISABLED:
    wm.setWifiEnabled(true);
    break;
case WifiManager.WIFI_STATE_ENABLING:
    break;
case WifiManager.WIFI_STATE_ENABLED:
    wm.setWifiEnabled(false);
    break;
}
```

常见异常

开关 WiFi 需要权限 android.permission.CHANGE_NETWORK_STATE 以及 android.permission.CHANGE_WIFI_STATE

（好累，感觉不会再贴图了）

第三节 获取系统信息

获取联系人信息

联系人信息的获取就需要我们在第 2 章讲到的四大组件中的 Content Provider。通过 ContentResolver 的对象来获取系统联系人的信息。

```
ArrayList<ListVo> list = new ArrayList<ListVo>();
//默认头像
Bitmap defaultIcon = BitmapFactory.decodeResource(
     getResources(),R.drawable.contacts);
ContentResolver resolver = getContentResolver();
//获取手机联系人
Cursor phoneCursor = resolver.query(Phone.CONTENT_URI,
     PHONES_PROJECTION,null,null,null);
if (phoneCursor != null) {
    while (phoneCursor.moveToNext()) {
        //得到手机号码
        String phoneNumber = phoneCursor.getString(1);
        phoneNumber = phoneNumber.replace(" ","");
        //当手机号码为空或者为空字段时跳过当前循环
        if (TextUtils.isEmpty(phoneNumber))
             continue;
        //得到联系人名称
        String contactName = phoneCursor.getString(0);

        Long contactId = phoneCursor.getLong(2);
        Long photoid = phoneCursor.getLong(3);
        //得到联系人头像 Bitamp
        Bitmap icon = defaultIcon;
        //photoid 大于 0 表示联系人有头像,如果没有给此人设置头像则给他一个默认的
        if(photoid > 0) {
            Uri uri = ContentUris.withAppendedId(
                    ContactsContract.Contacts.CONTENT_URI,contactId);
            InputStream input = ContactsContract.Contacts.
                    openContactPhotoInputStream(resolver,uri);
            icon = BitmapFactory.decodeStream(input);
        }
    }
    phoneCursor.close();
}
```

常见异常

读取联系人需要权限 android.permission.READ_CONTACTS,否则抛出异常:

```
Caused by:java.lang.SecurityException:Permission Denial:reading com.android.providers.
contacts.Contactsprovider2 uri content://com.android.contacts/data/phones from pid =
17602,uid = 10104 requires android.permission.READ_CONTACTS,or qrantUriPermission()
```

获取应用信息

我们知道,所有的应用的第一个 activity 都会包含这样一个 filter:

```
<intent-filter>
     <action android:name = "android.intent.action.MAIN"/>
     <category android:name = "android.intent.category.LAUNCHER"/>
</intent-filter>
```

也就是说，每个正常的有第一个 activity 的应用都会有这个 filter，因此我们可以利用这样一个特点来搜索系统中所有的应用：

```
/*
 * 查找符合条件的 activity
 * action 为 android.intent.action.MAIN
 * category 为 android.intent.category.LAUNCHER
 */
Intent mainIntent = new Intent(Intent.ACTION_MAIN,null);
mainIntent.addCategory(Intent.CATEGORY_LAUNCHER);
PackageManager packageManager = getPackageManager();
//查询 activity
List<ResolveInfo> list = packageManager.queryIntentActivities(mainIntent,0);
ArrayList<ListVo> appList = new ArrayList<ListVo>();
for(ResolveInfo res:list){
    //应用名称
    String label = res.loadLabel(packageManager).toString();
    //应用包名
    String packageName = res.activityInfo.packageName;
    //获取应用图标
    Drawable drawable = res.activityInfo.loadIcon(getPackageManager());
    BitmapDrawable bm = (BitmapDrawable)drawable;
    Bitmap icon = bm.getBitmap();
}
```

获取进程信息

获取运行中的应用

```
List<RunningAppProcessInfo> runningProcess = am.getRunningAppProcesses();
for (RunningAppProcessInfo amPro : runningProcess){

    //获得该进程占用的内存
    int[] myMempid = new int[] {amPro.pid};
    //此 MemoryInfo 位于 android.os.Debug.MemoryInfo 包中,用来统计进程的内存信息
    Debug.MemoryInfo[] memoryInfo = am.getProcessMemoryInfo(myMempid);

    //获取进程占内存信息形如 3.14MB
    double memSize = memoryInfo[0].dalvikPrivateDirty/1024.0;
    int temp = (int)(memSize * 100);
    memSize = temp/100.0;
    //获取进程名称
    String processName = amPro.processName;
}
```

获取系统可用内存：

```
ActivityManager.MemoryInfo memoryInfo = new ActivityManager.MemoryInfo();
am.getMemoryInfo(memoryInfo) ;
long memSize = memoryInfo.availMem ;
String memoLeft = "system idle memory:" + Formatter.formatFileSize(this,memSize);
```

杀死运行中的进程：

```
List<RunningAppProcessInfo> runningProcess = am.getRunningAppProcesses();
for (ActivityManager.RunningAppProcessInfo amPro : runningProcess){
    String processName = amPro.processName;
    am.restartPackage(processName);
}
```

调用闪光灯

闪光灯的操作是和相机的调用绑定在一起的，也就是说使用闪光灯必定会调用相机。相机功能通过 SurfaceView 来实现，相机中的预览画面本质上就是一个 SurfaceView，如：

```
public class TorchActivity extends Activity implements SurfaceHolder.Callback{

    private Camera camera;
    private SurfaceView mSurfaceView;
    private SurfaceHolder mSurfaceHolder;
    private boolean mIsSupported = false;

    @Override
    public void onCreate(Bundle savedInstanceState){
        super.onCreate(savedInstanceState);
        setContentView(R.layout.activity_torch);
        initTorch();
    }

    @Override
    public void onDestroy(){
        super.onDestroy();
    }

    private void initTorch(){
        if(getPackageManager().hasSystemFeature(
                        PackageManager.FEATURE_CAMERA_FLASH)) {
            try {
                mSurfaceView = (SurfaceView) findViewById(R.id.torch);
                mSurfaceHolder = mSurfaceView.getHolder();
                mSurfaceHolder.addCallback(this);
                mSurfaceHolder.setType(SurfaceHolder.SURFACE_TYPE_PUSH_BUFFERS);
                mIsSupported = true;
            } catch(Exception e) {
                e.printStackTrace();
            }
        }

        CheckBox toggle = (CheckBox) findViewById(R.id.toggle);
        toggle.setOnCheckedChangeListener(new OnCheckedChangeListener() {
```

```
            @Override
            public void onCheckedChanged(CompoundButton buttonView,
                                        boolean isChecked) {
                if(!mIsSupported){
                    Toast.makeText(TorchActivity.this,
                                "Flash light not supported",Toast.LENGTH_LONG);
                    return;
                }
                Camera.Parameters param = camera.getParameters();
                if(!isChecked){
                    //关闭闪光灯
                    param.setFlashMode(Camera.Parameters.FLASH_MODE_OFF);
                    camera.setParameters(param);
                    //停止相机预览
                    camera.stopPreview();
                }else{
                    //打开闪光灯
                    param.setFlashMode(Camera.Parameters.FLASH_MODE_TORCH);
                    camera.setParameters(param);
                    //开始相机预览
                    camera.startPreview();
                }
            }
        });
    }

    @Override
    public void surfaceChanged(SurfaceHolder arg0,int arg1,int arg2,int arg3)
    {
    }

    @Override
    public void surfaceCreated(SurfaceHolder holder) {
        try {
            if(camera == null){
                //打开相机
                camera = Camera.open();
            }
            //设置预览的画布
            camera.setPreviewDisplay(holder);
        } catch (Exception e) {
            e.printStackTrace();
        }
    }

    @Override
    public void surfaceDestroyed(SurfaceHolder arg0) {
        camera.release();
        camera = null;
    }
}
```

如果没有这个SurfaceView或者这个SurfaceView不可见(invisible),则获取不到camera的对象。当然,调用闪光灯需要权限android. permission. CAMERA。

第四节　功能实现

界面实现

本应用只有一个界面(长舒一口气),界面的布局也相对简单,如图6-8所示。此外,搜索结果栏的布局相对较为复杂,我们会在下面的一节中详细讲解。

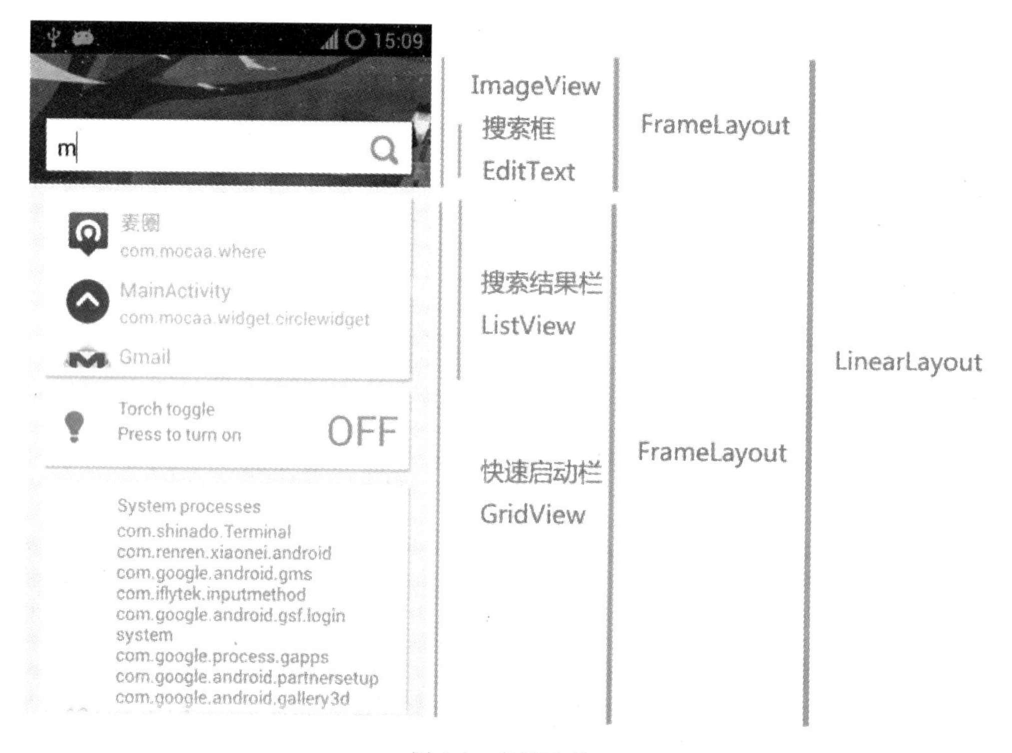

图6-8　布局示意

这个界面的框架就是:

```
<LinearLayout
    android:layout_width="fill_parent"
    android:layout_height="fill_parent"
    android:orientation="vertical">

    <FrameLayout
        android:layout_width="fill_parent"
        android:layout_height="wrap_content"
        >
        <SurfaceView
            android:layout_width="1dip"
```

```
                android:layout_height = "1dip"
                />
            < ImageView
                android:layout_width = "fill_parent"
                android:layout_height = "100dip"
                />
            < EditText
                android:layout_width = "300dip"
                android:layout_height = "45dip"
            />
        </FrameLayout >

        < FrameLayout
            android:layout_width = "fill_parent"
            android:layout_height = "fill_parent"
            >
            < GridView
                android:layout_width = "fill_parent"
                android:layout_height = "fill_parent"/>
            < com.shinado.Tearch.view.FlingListView
                android:layout_width = "300dip"
                android:layout_height = "150dip"
                />
        </FrameLayout >

    </LinearLayout >
```

值得注意的是，在本应该是 ListView 的快速启动栏在这里却使用了 GridView。这是因为 GridView 元素之间默认是有距离的，可以通过 android:verticalSpacing 进行控制，当 android:numColumns 为 1 时，这个 GridView 也就是 ListView 了。

有关 ListView 的细节我们已经在第 3 章和第 4 章有了介绍，在这里就不再赘述了。

系统主框架的实现

首先，使用一个 TextWatcher 来监听输入内容的改变：

```
        input = (EditText) findViewById(R.id.main_search_et);
        input.addTextChangedListener(mTextWatcher);
        //...
    }
    private TextWatcher mTextWatcher = new TextWatcher() {
            @Override
            public void beforeTextChanged(CharSequence s, int arg1, int arg2,
                    int arg3) {
            }

            @Override
            public void onTextChanged(CharSequence s, int arg1, int arg2,
```

```
                    int arg3) {
        }

        @Override
        public void afterTextChanged(Editable s) {
            //搜索
            results = doSearch(s.toString());
            if(results.size() == 0){
                //无结果,关闭搜索结果栏
                collapeList();
            }else{
                //展开搜索结果栏
                expandList(results);
            }
    }
};
```

开始搜索：

```
/*
 * 开始搜索
 * 返回搜索结果
 */
private ArrayList<ResultVo> doSearch(String key)
{
    ArrayList<ResultVo> list = new ArrayList<ResultVo>();
    if(key.equals("") || key == null){
        return list;
    }
    //让各个 function 进行搜索
    for(FunctionVo vo:functions){
        vo.getFunction().search(list,key);
    }

    return list;
}
```

搜索的具体实现依赖于 functions 这个对象。我们来看 functions 对象是怎么获取的：

```
/*
 * 载入搜索具体实现功能
 */
private void initFunction(){
        /*
         * 从配置文件中获取几个对象:
         *   String className;
        int type;
        String name;
        String img;
```

```
        String imgLeft;
        String imgRight;
          */
        functions = XMLUtil.getFunctions(this);
        for(FunctionVo entity:functions){
            ResultVo result = new ResultVo();
            result.setDisplayName(entity.getName());
            result.setType(entity.getType());
            try {
            Class cc = Class.forName(
                        PACKAGE_NAME_FUNCTION + entity.getClassName());
            //获得一个 IFuction 的实例化对象
            IFunction function = (IFunction)cc.newInstance();
            //初始化
            function.init(this,result);
            /*
              * 注入这个 IFuction
              * 见: getFunction
              */
            entity.setFunction(function);
            //放入 map 以便获取
            functionMap.put(entity.getType(),function);
        } catch (InstantiationException e) {
            e.printStackTrace();
        } catch (IllegalAccessException e) {
            e.printStackTrace();
        } catch (ClassNotFoundException e) {
            e.printStackTrace();
        }
    }
}
```

IFunction 是一个抽象类：

```
public abstract class IFunction {

    protected ResultVo result;
    protected Context context;

    public void init(Context context,ResultVo result){
        this.context = context;
        this.result = result;
    }
    /**
      * 单击结果栏
      * @param result 执行的结果信息
      */
    public abstract void function(ResultVo result);
```

```
    /**
     * 向左滑动结果栏
     * @param result 执行的结果信息
     */
    public abstract void slideLeft(ResultVo result);
    /**
     * 向右滑动结果栏
     * @param result 执行的结果信息
     */
    public abstract void slideRight(ResultVo result);
    /**
     * 搜索
     * @param list 搜索的结果
     * @param key  搜索关键字
     */
    public abstract void search(ArrayList<ResultVo> list,String key);
    /**
     * 刷新
     */
    public abstract void refresh();

    public ResultVo getResult() {
        return result;
    }
    public void setResult(ResultVo result) {
        this.result = result;
    }
}
```

再来看搜索联系人功能搜索的实现伪代码：

```
@Override
public void search(ArrayList<ResultVo> list,String key) {
    /*
      * 搜索联系人信息
      * 在所有联系人的信息中匹配,将匹配到的联系人数据放到 list 中
      * 是否有包含 key 的数据
      * list 搜索到的结果数据
      */
}
```

这里有一个非常重要的实体类 ResultVo,它包含了几个属性：

```
public class ResultVo implements Cloneable{
    //显示的图片
    private Bitmap img;
    /*
     * 用于匹配的名称
     * 当 displayName 是中文时,此对象为 displayName 的拼音
     */
```

```
    private String name;
    //用于显示的名称
    private String displayName;
    /*
     * 用于执行的信息,
     * 如包名、号码等
     */
    private String value;
    private int type;
    public static final int TYPE_APP = 1;
    public static final int TYPE_CONTACT = 2;
    public static final int TYPE_ORDER = 3;
}
```

例如,输入 lao 时,会出现老邓的联系人信息,此时这个 ResultVo 的各个属性值如图 6-9 所示。

图 6-9 ResultVo 属性示意

当搜索有结果时,调用函数:

```
private void expandList(ArrayList<ResultVo> results){
    if(resultView.getVisibility() != View.VISIBLE){
            resultView.setVisibility(View.VISIBLE);
            resultView.startAnimation(anim_in);
    }
    resultAdapter.setList(results);
    resultAdapter.notifyDataSetChanged();
}
```

resultView 跟 resultAdapter 的初始化:

```
resultView = (FlingListView) findViewById(R.id.main_result_list);
resultAdapter = new ResultListAdapter(this,functions);
resultView.setAdapter(resultAdapter);
```

FlingListView 这个类是自定义的可滑动的 ListView,跟第 4 章介绍的滑动 ListView 有点类似,这个类会在接下来的一节具体讲解。ResultListAdapter 类的伪代码如下:

```
public class ResultListAdapter extends BaseAdapter{

    private Context context;
    private ArrayList<ResultVo> list;
    private HashMap<Integer,ImgsVo> imgMap = new HashMap<Integer,ImgsVo>();
```

```
    public void setList(ArrayList<ResultVo> list) {
        this.list = list;
    }

    /**
     * @param context
     * @param functions 用于初始化图标
     */
    public ResultListAdapter(Context context,ArrayList<FunctionVo> functions) {
        this.context = context;
        initBitmap(functions);
    }

    /**
     * 初始化图标
     * @param functions
     */
    private void initBitmap(ArrayList<FunctionVo> functions){
        //从 functions 中初始化图标
    }

    @Override
    public int getCount() {
        if(list != null){
            return list.size();
        }
        return 0;
    }

    @Override
    public Object getItem(int position) {
        return null;
    }

    @Override
    public long getItemId(int position) {
        return 0;
    }

    @Override
    public View getView(int position,View convertView,ViewGroup parent) {
        //设置视图内容

        /**
         * 重要: 设置 id 用于获取正确的位置
         */
        convertView.setId(position);
        return convertView;
    }
}
```

接下来，设置单击结果栏的某个元素的监听器：

```
    resultView.setOnItemClickListener(new OnItemClickListener() {
        @Override
        public void onItemClick(AdapterView<?> arg0,View arg1,int position,
                long arg3) {
            action(position);
        }
    });
    //此处省去 n 行代码
}
private void action(int position){
        if(position >= results.size()){
            return;
        }
        ResultVo result = results.get(position);
        //根据 type 来获取 Function 的具体实现对象
        IFunction func = functionMap.get(result.getType());
        if(func != null){
            //执行
            func.function(result);
        }
    //清空搜索栏中的内容
    input.setText("");
}
```

联系人功能的具体实现：

```
@Override
public void function(ResultVo rs) {
    //拨号
    String body = rs.getValue();
    if(!body.equals("")){
        Intent myIntentDial = new Intent(
                        Intent.ACTION_CALL,Uri.parse("tel:" + body));
        context.startActivity(myIntentDial);
    }
}
```

接下来设置结果栏左右滑动的监听器(这个监听器是自己实现的，接下来一节中会讲解)：

```
    resultView.setOnItemMovingListener(new OnItemMovingListener() {
        @Override
        public void onItemMoving(int absolutePosition,int movingDirection) {
            action(absolutePosition,movingDirection);
        }
    });
    //此处省去 n 行代码
}
```

```
private void action(int position, int movingDirection){
        ResultVo result = results.get(position);
        IFunction function = functionMap.get(result.getType());
        if(function != null){
            switch(movingDirection){
            case FlingListView.DIRECTION_LEFT:
                function.slideLeft(result);
                break;
            case FlingListView.DIRECTION_RIGHT:
                function.slideRight(result);
                break;
            }
    }
        input.setText("");
}
```

联系人功能的具体实现：

```
@Override
public void slideLeft(ResultVo rs) {
    //调用发送短信界面
    String body = rs.getValue();
    Intent intent = new Intent();
    intent.setAction(Intent.ACTION_SENDTO);
    //需要发短信的号码
    intent.setData(Uri.parse("smsto:" + body));
    intent.putExtra("sms_body",num);
    context.startActivity(intent);
    num = "";
}

@Override
public void slideRight(ResultVo rs) {
    //复制联系人信息
    num = rs.getDisplayName() + " " + rs.getValue();
}
```

到这里，搜索功能就(慢！还有收起搜索结果栏的函数)……好吧，其实也很简单，相信大家可以很轻松读懂：

```
private void collapeList(){
        if(resultView.getVisibility() != View.GONE){
            resultView.startAnimation(anim_out);
            new Thread(){
                public void run(){
                    try {
                        sleep(anim_out.getDuration());
                    } catch (InterruptedException e) {
```

```
                    e.printStackTrace();
                }
                handler.sendEmptyMessage(0);
            }
        }.start();
    }
}
```

快速启动的实现

首先,初始化快速启动 GridView:

```
GridView gridView = (GridView) findViewById(R.id.main_gridview);
//获取快速启动
launches = getLaunches();
gridAdapter = new GridViewAdapter(this, launches);
gridView.setAdapter(gridAdapter);
for(IPlugin plugin:plugins){
    plugin.setAdapter(gridAdapter);
}
//设置事件监听
gridView.setOnItemClickListener(new OnItemClickListener() {
    @Override
    public void onItemClick(AdapterView<?> arg0, View view, int position,
            long arg3) {
        if(view.isEnabled()){
            for(IPlugin plugin:plugins){
                if(position == plugin.getPlugin().getIndex()){
                    plugin.toggle(IPlugin.FLAG_CLICK);
                }
            }
        }
    }
});
gridView.setOnItemLongClickListener(new OnItemLongClickListener() {
    @Override
    public boolean onItemLongClick(AdapterView<?> arg0, View view,
            int position, long arg3) {
        if(view.isEnabled()){
            for(IPlugin plugin:plugins){
                if(position == plugin.getPlugin().getIndex()){
                    plugin.toggle(IPlugin.FLAG_LONG_CLICK);
                }
            }
        }
        return false;
    }

});
```

这里的 getLaunches()函数为：

```
private ArrayList<LaunchVo> getLaunches(){
    ArrayList<LaunchVo> list = new ArrayList<LaunchVo>();
    for(IPlugin plugin:plugins){
        list.add(plugin.getPlugin().getLaunch());
            int index = list.size() - 1;
            //设置 index
            plugin.getPlugin().setIndex(index);
    }
        return list;
}
```

plugins 从配置文件中读取：

```
plugins = new ArrayList<IPlugin>();
/*
  * 从配置文件中获取几个对象:
  *    String className;
      int id;
    String[] actions;
    String image;
  */
ArrayList<XMLPlugin> list = XMLUtil.getPlugins(this);
for(XMLPlugin entity:list){
    try {
        String className = entity.getClassName();
        Class cc = Class.forName(PACKAGE_NAME_PLUGIN + className);
        IPlugin plugin = (IPlugin)cc.newInstance();
        plugin.init(this,
                entity.getId(),
                entity.getActions(),
                ResourcesUtil.getImage(entity.getImage())
            );
            plugins.add(plugin);
        } catch (InstantiationException e) {
            e.printStackTrace();
        } catch (IllegalAccessException e) {
            e.printStackTrace();
        } catch (ClassNotFoundException e) {
            e.printStackTrace();
        }
}
```

同样，IPlugin 也是一个抽象类，继承自它的类需要实现：初始化显示信息、接收广播并响应、响应长按或者单击事件等：

```
public abstract class IPlugin {

    private PluginVo plugin;
```

```
    protected Context context;
    protected GridViewAdapter adapter;

    public void setAdapter(GridViewAdapter adapter) {
        this.adapter = adapter;
    }

    public void init(Context context, int id, String[] actions, int imgId){
        this.context = context;
        LaunchVo launch = new LaunchVo(
                    imgId, getInitText(), getInitDetail_1(), getInitDetail_2());
        PluginVo vo = new PluginVo(id, 0, actions, launch);
        this.plugin = vo;
    }

    /**
     * 接收到广播时的操作
     * @param context
     * @param intent
     */
    public abstract void onReceive(Context context, Intent intent);

    /**
     * 单击或长按的操作
     * @param flag 单击或者长按
     */
    public abstract void toggle(int flag);
    /**
     * 初始化时 text 的内容
     */
    public abstract String getInitText();
    /**
     * 初始化时 detail_2 的内容
     */
    public abstract String getInitDetail_1();
    /**
     * 初始化时 detail_1 的内容
     */
    public abstract String getInitDetail_2();

    public PluginVo getPlugin() {
        return plugin;
    }

    public void setPlugin(PluginVo plugin) {
        this.plugin = plugin;
    }
}
```

IPlugin 有一个比较纠结的对象 PluginVo：

```
public class PluginVo {
    //id,不解释
    private int id;
    //在快速启动界面中的序号
    private int index;
    //接收的广播的 action
    private String[] actions;
    //显示的视图实体
    private LaunchVo launch;
}
```

LaunchVo 的属性以及与界面相对应的解释如图 6-10 所示。

```
public class LaunchVo {

    private int imgId;
    private String text;
    private String detail_1;
    private String detail_2;
    private boolean enable = true;
}
```

imageId(图标资源id)
Torch toggle
Press to turn on
OFF
detail_1
detail_2
text

图 6-10　LounchVo 属性示意

以 WiFi 设置的实现为例：

```
public class WifiPlugin extends IPlugin{

    @Override
    public void onReceive(Context context, Intent intent) {
        int index = getPlugin().getIndex();
        LaunchVo launch = adapter.getList().get(index);
        if (WifiManager.WIFI_STATE_CHANGED_ACTION.equals(intent.getAction()))
        {
                //获取 WiFi 状态
              int wifiState = intent.getIntExtra(WifiManager.EXTRA_WIFI_STATE,0);
              switch (wifiState)
                {
                //WiFi 已打开
                case WifiManager.WIFI_STATE_ENABLED:
                    launch.setEnable(true);
                    launch.setText("ON");
```

```
                launch.setDetail_2("Press to turn it off");
                break;
            //WiFi 已关闭
            case WifiManager.WIFI_STATE_DISABLED:
                launch.setEnable(true);
                launch.setText("OFF");
                launch.setDetail_2("Press to turn it on");
                break;
            //WiFi 正在打开
            case WifiManager.WIFI_STATE_ENABLING:
            //WiFi 正在关闭
            case WifiManager.WIFI_STATE_DISABLING:
                launch.setEnable(false);
                break;
            }
        }
        adapter.notifyDataSetChanged();
    }

    @Override
    public void toggle(int flag) {
        WifiManager wm = (WifiManager)context.getSystemService(
                Context.WIFI_SERVICE);
        if(flag == FLAG_CLICK){
            //单击,开关 WiFi
            switch(wm.getWifiState()){
        case WifiManager.WIFI_STATE_DISABLED:
                wm.setWifiEnabled(true);
                break;
            case WifiManager.WIFI_STATE_ENABLED:
                wm.setWifiEnabled(false);
                break;
            }
            }else{
            //长按,打开 WiFi 设置界面
            context.startActivity(new Intent(
                    android.provider.Settings.ACTION_WIFI_SETTINGS));
        }
    }

    private boolean isWifiOn(){
        WifiManager wm = (WifiManager)context.getSystemService(
                        Context.WIFI_SERVICE);
        switch(wm.getWifiState()){
        case WifiManager.WIFI_STATE_ENABLED:
            return true;
        default:
                return false;
            }
    }
```

```
    @Override
    public String getInitText() {
        return isWifiOn()? "ON":"OFF";
    }

    @Override
    public String getInitDetail_1() {
        return "Wifi status";
    }

    @Override
    public String getInitDetail_2() {
        return "Press to turn " + (isWifiOn()? "off":"on");
    }
}
```

滑动列表的实现

这个应用中的滑动列表的实现与第 4 章的滑动列表有点相似，但也有不同的地方：当列表的元素向右(左)滑动时，元素左(右)边会弹出一个提示图标，并且在滑动的过程中保持位置不变，如图 6-11 所示。

图 6-11　滑动列表

这个布局的伪代码如下：

```
<RelativeLayout>

    <!-- 左图标,默认靠左 -->
    <ImageView
        android:id="@+id/layout_list_drawable_left"
        />
    <!-- 显示界面 -->
    <LinearLayout>
    </LinearLayout>

    <!-- 右图标
         layout_alignParentRight -->
    <ImageView
        android:id="@+id/layout_list_drawable_right"
        android:layout_alignParentRight="true"
        />

</RelativeLayout>
```

这个功能实现依靠于 margin_left 与 margin_right 的动态设置。当整个元素往右滑动时，改变 layout_list_drawable_left 这个 ImageView 的 margin_left 的值，margin_left 的值正是元素移动的距离如图 6-12 所示；当整个元素往左滑动时同理。

图 6-12　列表滑动的原理

```
private boolean move(int offset){
    if(mChildView == null){
        //空白区域
        if(mChildPosition >= FlingListView.this.getChildCount()){
            mIsToSlide = false;
            return true;
        }
        //获取子元素
        mChildView = getChildAt(mChildPosition);
        mDrawableLeft = (ImageView)mChildView.findViewById(
                R.id.layout_list_drawable_left);
        mDrawableRight = (ImageView)mChildView.findViewById(
                R.id.layout_list_drawable_right);
    }
    mChildView.scrollTo(offset,0);

    /*
     * 向右滑动,出现左边的图标
     */
    if(offset < 0){
        RelativeLayout.LayoutParams params =
                (RelativeLayout.LayoutParams) mDrawableLeft.getLayoutParams();
        if( - offset > params.width){
            if(mDrawableLeft.getVisibility() != View.VISIBLE){
                //出现右边图标
                movingDirection = DIRECTION_RIGHT;
                mDrawableLeft.setVisibility(View.VISIBLE);
                mDrawableLeft.startAnimation(anim);
            }
```

```
        }else{
            if(mDrawableLeft.getVisibility() != View.GONE){
                movingDirection = DIRECTION_NONE;
                mDrawableLeft.setVisibility(View.GONE);
            }
        }
        /*
         * 标题要长长长长长长长长长长长长长长长长长长长长长长长长长长长长长长长长长长长长
         * 实现的关键
         */
        params.leftMargin = offset;
        mDrawableLeft.setLayoutParams(params);
    }
    //leftwards
    else{
        RelativeLayout.LayoutParams params =
                (RelativeLayout.LayoutParams) mDrawableRight.getLayoutParams();
        if(offset > params.width){
            if(mDrawableRight.getVisibility() != View.VISIBLE){
                //出现左边图标
                movingDirection = DIRECTION_LEFT;
                mDrawableRight.setVisibility(View.VISIBLE);
                mDrawableRight.startAnimation(anim);
            }
        }else{
            if(mDrawableRight.getVisibility() != View.GONE){
                movingDirection = DIRECTION_NONE;
                mDrawableRight.setVisibility(View.GONE);
            }
        }
        /*
         * 标题要长长长长长长长长长长长长长长长长长长长长长长长长长长长长长长长长长长长长
         * 实现的关键
         */
        params.rightMargin = - offset;
        mDrawableRight.setLayoutParams(params);
    }

    return false;
}
```

第七章　MyWhere

地图＋增强现实

清单

Demo 代码：

\demo\Demo_Where

实例代码：

\source_codes\MyWhere

Target SDK：

Android 2.2

第一节　产 品 介 绍

这一章要介绍的实例不能算是一个实际的产品，因为是从一个大项目中抽离出来的，只有几个简单的小功能。但是这一章会详细地讲解增强现实的实现，并且对其中的几个难点进行详细的分析。

这个产品借鉴(模仿)了 wikitude 的增强现实导航功能，如图 7-1 所示。

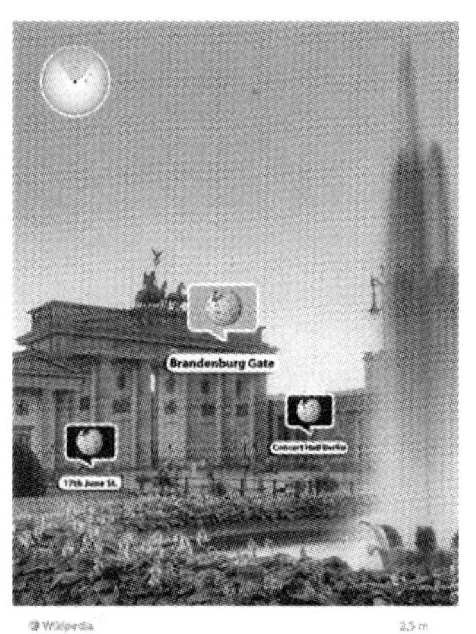

图 7-1　wikitude 应用

需求分析

定位、搜索周边

通过百度提供的 API，搜索周边，如图 7-2 所示。

增强现实显示

将搜索到的消息通过增强现实的方式展示出来，如图 7-3 所示。

图 7-2　搜索周边功能

图 7-3　增强现实显示

兴趣点

移动、缩放兴趣点；借助百度 API，单击兴趣点进入详情页面。

拍照

保存当前画面。

第二节　地 图 开 发

在开发地图应用时，我们一般不使用 Android 自带的 Google 地图，原因有多个。首先是谷歌的服务在中国内地不稳定；其次，有许多国产的 ROM 去掉了 Google 地图的库，在这样的机器上是运行不起来的。因此，本应用使用百度地图来搭建。

进入网站 http://developer. baidu. com/map/sdk-android. htm，在这里有非常丰富的开发示例，在这里我就不多赘述了。

第三节　传感器开发

传感器种类

Android 2. 3 支持的传感器有 11 种，如表 7-1 所示。

表 7-1　传感器

传感器名称	含　义
ACCELEROMETER	加速度感应器
GRAVITY	重力感应器
GYROSCOPE	陀螺仪感应器
LIGHT	光线感应器
LINEAR_ACCELERATION	线性加速度感应器
MAGNETIC_FIELD	磁力感应器
ORIENTATION	方位感应器
TYPE_PRESSURE	压强感应器
TYPE_PROXIMITY	距离感应器
ROTATION_VECTOR	旋转向量感应器
EMPERATURE	温度感应器

重要的感应器为加速度感应器、方位感应器以及陀螺仪感应器。在这一节只介绍这三个感应器的知识。这三个感应器都有三个数据：X、Y、Z。为了讲解方便，假定手机的坐标系如图 7-4 所示。

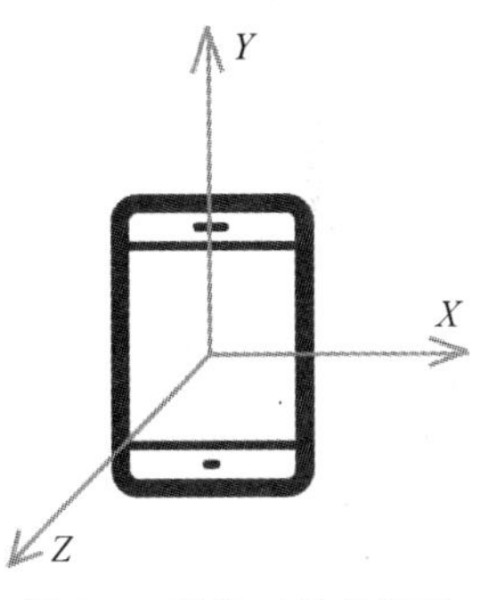

图 7-4　手机三轴坐标系

加速度感应器

顾名思义，加速度感应器就是测量各个方向的加速度。由于重力的存在，当手机平放时，会有一个方向的值不为 0，而其他两个方向值为 0。手机的摆放位置及 X、Y、Z 的值如表 7-2 所示。

表 7-2　手机摆放位置与 X、Y、Z 的值

摆放位置	X	Y	Z
平放朝上(Z 正半轴方向)	0	0	9.8
平放朝下(Z 负半轴方向)	0	0	−9.8
直立朝下(Y 负半轴方向)	0	−9.8	0
直立朝上(Y 正半轴方向)	0	9.8	0
屏幕朝左(Z 负半轴方向)	9.8	0	0
屏幕朝右(Z 正半轴方向)	−9.8	0	0

方向感应器

方向感应器的原理是利用磁极感应来测出手机的朝向。其中，X 值指示方向，Y 值指示手机 Y 轴相对于地面的旋转量，Z 表示手机 X 轴相对于地面的旋转量，X 值指示方向。

将手机平放于水平面上，Y 跟 Z 的值都为 0，而 X 的值指向了手机的方向。此时，若手机的 Y 轴正半轴指向正北方，则 X 的数值为 0(360)，西为 270，南不为 180，东为 90。

将手机绕着 X 轴旋转，当 Y 轴正半轴指向天空时，Y 值为 −90，指向地面为 90，而当 Z 轴正半轴(屏幕)指向地面时，Y 值为 180(−180)。

将手机绕着 Z 轴旋转，当 X 正半轴指向地面时，Z 值为 −90，指向天空为 90，而当 Z 轴正半轴(屏幕)指向地面时，Z 值为 0。

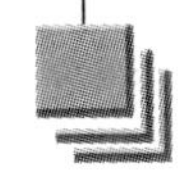

陀螺仪感应器

陀螺仪感应器记录的是手机在任何一维上的旋转变化量。所以当手机绝对静止的时候，X、Y、Z 的值都为 0。

当手机绕着 X 轴旋转时，X 值开始变化，数值的绝对值指示了旋转速度。

当手机绕着 Y 轴旋转时，Y 值开始变化，数值的绝对值指示了旋转速度。

当手机绕着 Z 轴旋转时，Z 值开始变化，数值的绝对值指示了旋转速度。

传感器使用

传感器的使用很简单，注册一个就行：

```
public class SensorDetailActivity extends Activity {

    private SensorManager sensorManager;

    @Override
    public void onCreate(Bundle savedInstanceState){
        super.onCreate(savedInstanceState);
        setContentView(R.layout.activity_sensor_detail);

        int type = Sensor.TYPE_ACCELEROMETER
        //获取传感器服务类
        sensorManager = (SensorManager) getSystemService(SENSOR_SERVICE);
        Sensor sensor_orientation = sensorManager.getDefaultSensor(type);
        /*
         * 注册传感器
         * sensorListener 监听器
         * sensor_orientation 传感器类型
         * SensorManager.SENSOR_DELAY_UI 传感器精度，有
         * SensorManager.SENSOR_DELAY_UI
         * SensorManager.SENSOR_DELAY_NORMAL
         * SensorManager.SENSOR_DELAY_GAME
         * SensorManager.SENSOR_DELAY_FASTEST
         * 精度依次增大
         */
         sensorManager.registerListener(sensorListener,
            sensor_orientation,
            SensorManager.SENSOR_DELAY_UI);
    }

    private SensorEventListener sensorListener = new SensorEventListener(){
        @Override
        //可以得到传感器实时测量出来的变化值
        public void onSensorChanged(SensorEvent event) {
            float x = event.values[SensorManager.DATA_X];
            float y = event.values[SensorManager.DATA_Y];
            float z = event.values[SensorManager.DATA_Z];
        }
```

```
        @Override
        public void onAccuracyChanged(Sensor arg0, int arg1) {
        }
    };
    @Override
    public void onDestroy(){
        super.onDestroy();
        //记得要注销哦
        sensorManager.unregisterListener(sensorListener);
    }

}
```

需要注意的是，在网上搜索的时候可能会看到这样注册的方式：

```
sensorManager.registerListener(new SensorListener(){
    @Override
    public void onAccuracyChanged(int arg0, int arg1) {
    }
    @Override
    public void onSensorChanged(int arg0, float[] arg1) {
    }
},
Sensor.TYPE_ACCELEROMETER,
SensorManager.SENSOR_DELAY_UI);
```

这个方法已经被 Google 废弃了，最好不要使用，这个方法在老版本的设备上可以使用，但是在新设备(如小米 2)上就无法正常获取数据。

第四节 相机开发

相机的预览画面本质上是一个 SurfaceView，通过使用 Camera 封装的方法，来获取相机预览、拍照等相机功能。

相机画面预览

我们在前面一章提到过相机的功能，在这一节我们会详细讲解相机预览画面的使用。首先，很常规的开局：

```
public class CameraView extends SurfaceView implements SurfaceHolder.Callback{

    public CameraView(Context context, AttributeSet attrs) {
        super(context, attrs);
        this.mContext = context;
        mHolder = this.getHolder();
        mHolder.setType(SurfaceHolder.SURFACE_TYPE_PUSH_BUFFERS);
        mHolder.addCallback(this);
        //设置透明图层
        this.setZOrderOnTop(false);
    }
```

在重写的 surfaceCreated 方法中，加入细节设置。

第一步，获取 camera 对象。

```
mCamera = Camera.open();
```

第二步，调整相机方位。由于 Android 对相机的设定默认是横向的，我们需要把相机调正过来(略显奇怪的设定)。注意，此方法是 2.2 以上的 SDK 才有的：

```
mCamera.setDisplayOrientation(90);
```

第三步，获取设备支持的相机预览大小以及照片大小。这一步比较纠结，因为有些设备不支持正常的照片大小如 480×800。如果设置了设备不支持的大小就会出现错误：

```
java.lang.RuntimeException:setParameters failed
at android.hardware.Camera.native_setParameters(Native Method)
at android.hardware.Camera.setParameters(camera.java:1490)
at com.example.demo_where.CameraView.surfaceCreated (CameraView.java:89)
```

因此我写了一个方法，通过和设备屏幕大小的比较来获取合适的预览大小：

```
/**
 * @param display 屏幕显示的大小
 * @param supportedPreviewSizes 设备支持的预览/照片大小
 * @return
 */
private Size getSuitableParameter(Display display,
            List<Size> supportedPreviewSizes){
    //获取设备支持的大小
    int min = 100000;
    int index = 0;
    for(int i = 0; i < supportedPreviewSizes.size(); i++){
        Size size = supportedPreviewSizes.get(i);
        //如果支持的长宽跟屏幕长宽有一个相等,就是它了
        if(size.width == display.getHeight()){
            return size;
        }
        if(size.height == display.getWidth()){
            return size;
        }
        //否则,取两者差值最小的
        int gapHeight = Math.abs(size.height - display.getWidth());
        int gapWidth = Math.abs(size.width - display.getHeight());
        if(gapHeight + gapWidth < min){
            min = gapHeight + gapWidth;
            index = i;
        }
    }
    return supportedPreviewSizes.get(index);
}
```

设置相机预览的大小以及照片大小：

```
//获取窗口显示的大小,用于比较
WindowManager wm = (WindowManager) mContext
        .getSystemService(Context.WINDOW_SERVICE);
Display display = wm.getDefaultDisplay();
Camera.Parameters parameters = mCamera.getParameters();
//获取合适的、手机支持的预览大小
Size previewSize = getSuitableParameter(display,
        parameters.getSupportedPreviewSizes());
//设置预览大小
parameters.setPreviewSize(previewSize.width,previewSize.height);
//获取合适的、手机支持的预览拍照图片大小
Size picturSize = getSuitableParameter(display,
        parameters.getSupportedPictureSizes());
//设置拍照图片大小
parameters.setPictureSize(picturSize.width,picturSize.height);
```

第四步,设置相机帧数。同样的老问题,需要获取设备支持的相机帧数进行设置：

```
//设置相机帧数
parameters.setPreviewFrameRate(getPreviewRate(parameters));
...
private int getPreviewRate(Camera.Parameters parameters){
    List<Integer> list = parameters.getSupportedPreviewFrameRates();
    if(list.size() > 0){
        return list.get(list.size()/2);
    }else{
        return -1;
    }
}
```

最后,设置其他参数,开始预览：

```
//设置相机成像格式
parameters.setPictureFormat(PixelFormat.JPEG);
//设置相片质量
parameters.setJpegQuality(85);
mCamera.setParameters(parameters);
/*
 *关键,设置预览的 surface
 */
mCamera.setPreviewDisplay(mHolder);
//开始预览
mCamera.startPreview();
mIsPreview = true;
```

在画布销毁的时候释放 camera 对象：

```
@Override
public void surfaceDestroyed(SurfaceHolder holder) {
    if(mCamera!= null){
        if(mIsPreview){
            mCamera.stopPreview();
        }
        mCamera.release();
    }else{
        Log.e("camera","null");
    }
}
```

最后要记得加上使用相机的权限：

```
<uses-permission android:name="android.permission.CAMERA"/>
```

拍照并保存

调用 Camera 的 takePic 方法，提供一个回调函数设置拍照后的处理：

```
public void takePic(){
    if(mCamera == null){
        Log.e("camera","null");
        return;
    }
    mCamera.takePicture(null,null,picCallback);
    //拍照自动停止预览
    mIsPreview = false;
}
private static final String AR_PIC_PATH = "/sdcard/demo_where/";
private PictureCallback picCallback = new PictureCallback(){
    @Override
    public void onPictureTaken(final byte[] data,Camera camera) {
        if(mpd == null){
            //设置等待对话框
             mpd = new ProgressDialog(mContext);
             mpd.setProgressStyle(ProgressDialog.STYLE_SPINNER);
             mpd.setIndeterminate(false);
             mpd.setCancelable(true);
             mpd.setTitle("正在保存,请稍候");
        }
        mpd.show();

        new Thread(){
             public void run(){
                 //获取拍照的 Bitmap 对象
                 Bitmap photo = BitmapFactory.decodeByteArray(
                         data,0,data.length,null);
                 /*
```

```
                * 保存照片
                * AR_PIC_PATH 路径
                * getTimeInSec() 文件名,为了确保唯一,以年月日时分秒来命名
                */
                boolean flag = saveBitmap(AR_PIC_PATH,getTimeInSec(),photo);
                if(flag){
                    mHandler.sendEmptyMessage(1);
                }else{
                    mHandler.sendEmptyMessage(-1);
                }
                continuePreview();
            }
        }.start();
    }
};
```

保存图片的方法依然是使用I/O来实现：

```
private boolean saveBitmap(String url,String picName,Bitmap bitmap){
    try{
            File newfolder = new File(url);
            //如果不存在该路径,创建一个文件夹
            if(!(newfolder.exists())&&!(newfolder.isDirectory())){
                newfolder.mkdirs();
            }
            url = url + picName + ".jpg";
            File f = new File(url);
            f.createNewFile();
            FileOutputStream fOut = null;
            fOut = new FileOutputStream(f);
            //保存图片
            boolean flag = bitmap.compress(Bitmap.CompressFormat.JPEG,100,fOut);
            fOut.flush();
            fOut.close();
            return flag;
    }catch(Exception e){
            e.printStackTrace();
    }
    return false;
}
```

由于保存图片是对存储卡进行写入的操作，因此需要权限：

```
<uses-permission android:name="android.permission.WRITE_EXTERNAL_STORAGE"/>
```

第五节　Canvas 绘图

在 SurfaceView 的世界里，所有的元素都是通过 Canvas 绘制上去的，所以不要指望有什么 ImageView、Button、TextView 了，需要自己绘制文本、图形，甚至按钮。所以在这一节我们要学习 Canvas 绘图，将图片、文本等元素通过裁剪、修边、粘合等方式在 Canvas 上绘图。

图片粘合

首先，绘制一个 canvas 背景。这个 Bitmap 必须是透明的。

```
/*
 * 绘制一个 mutable 的位图作为 canvas 的背景
 * 长: bgHeight 宽: bgWidth
 */
Bitmap newbmp = Bitmap.createBitmap(bgWidth,bgHeight,Config.ARGB_8888);
Canvas cv = new Canvas(newbmp);
```

接下来，在这个画布上绘制你想要的位图：

```
cv.drawBitmap(background,0,0,null);
cv.drawBitmap(foreground,offset.x,offset.y,null);
```

绘制完成之后，这张 Canvas 的位图 newbmp 就是我们需要的粘合效果。完整的函数为：

```
public Bitmap combineBitmap(Bitmap background,Bitmap foreground,Point offset) {
    if(background == null || foreground == null) {
        return null;
    }
    if(offset == null){
        offset = new Point(0,0);
    }
    int bgWidth = background.getWidth();
    int bgHeight = background.getHeight();

    Bitmap newbmp = Bitmap.createBitmap(bgWidth,bgHeight,Config.ARGB_8888);
    Canvas cv = new Canvas(newbmp);

    cv.drawBitmap(background,0,0,null);
    cv.drawBitmap(foreground,offset.x,offset.y,null);
    return newbmp;
}
```

图片剪切

首先，准备好一张你需要剪切的形状的图片，如图 7-5 所示，透明部分表示需要裁剪的部分。

图 7-5 剪切的形状

接下来的代码与第一个没有什么区别，但是加上了一支画笔：

```
Paint paint = new Paint();
paint.setXfermode(new PorterDuffXfermode(Mode.SRC_IN));
```

这段代码的意思是设置两图相交时的模式，这个模式表示后面的图将会覆盖前面的图，而前面图透明的部分将会消失。完整的代码如下：

```
public Bitmap clipBitmap(Bitmap bitmap,Bitmap cover){
    Bitmap output = Bitmap.createBitmap(bitmap.getWidth(),bitmap
         .getHeight(),Config.ARGB_8888);
    Canvas canvas = new Canvas(output);

    canvas.drawBitmap(cover,0,0,null);
    /*
     * 关键,设置两图相交时的绘图模式
     */
    Paint paint = new Paint();
    paint.setXfermode(new PorterDuffXfermode(Mode.SRC_IN));
    canvas.drawBitmap(bitmap,0,0,paint);

    return output;
}
```

绘制文本

Android 的文本绘制使用了 canvas.drwaText 的方法，代码示例如下：

```
public Bitmap drawText(Bitmap photo,String content){

    Bitmap output = Bitmap.createBitmap(photo.getWidth(),photo.getHeight(),
             Config.ARGB_8888);
    Canvas canvas = new Canvas(output);

    Paint contentPaint = new Paint();
    contentPaint.setARGB(255,255,255,255);
    contentPaint.setTextSize(23);
    canvas.drawBitmap(photo,0,0,null);
```

```
    canvas.drawText(content,8,35,contentPaint);
    return output;
}
```

扩展 9PNG 图片

如果是需要进行拉伸的 9PNG 格式图片，则需要通过代码实现：

```
public Bitmap draw9png(Bitmap img9png,float width){
    NinePatch np = new NinePatch(img9png,img9png.getNinePatchChunk(),null);
    Bitmap src = Bitmap.createBitmap(
            (int) (img9png.getWidth() * width),img9png.getHeight(),
            Config.ARGB_8888);
    Canvas c = new Canvas(src);
    c.drawBitmap(src,0,0,null);
    np.draw(c,new Rect(0,0,
            (int) (img9png.getWidth() * width),img9png.getHeight()));
    return src;
}
```

第六节 功能实现

本实例的重点只有一个 Activity，增强现实界面的 Activity，在本节里面我也是围绕这个 Activity 来讲解的。

增强现实布局的实现

在这个界面中包含了多个图层，从内到外分别为相机图层、兴趣点图层、控件图层以及雷达图层，如图 7-6 所示。

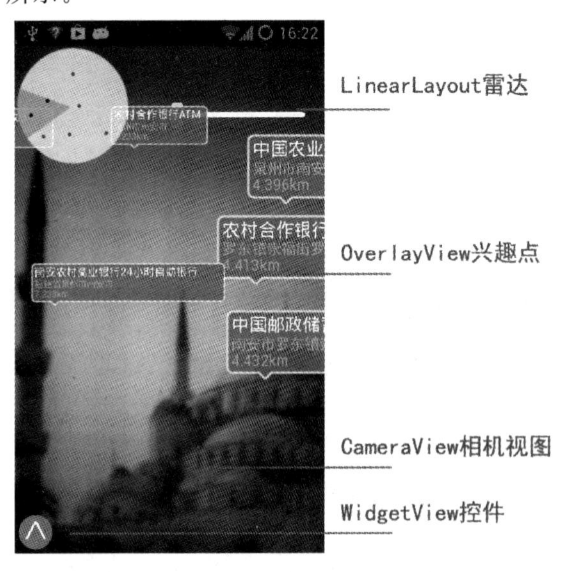

图 7-6　布局示意

整个布局的框架为：

```
<RelativeLayout
    >
    <com.shinado.mywhere.view.CameraView
        android:layout_width = "fill_parent"
        android:layout_height = "fill_parent"
    />
    <com.mocaa.where.overlay.OverlayView
        android:layout_width = "fill_parent"
        android:layout_height = "fill_parent"
    />
    <com.shinado.wherewidget.WidgetView
        android:layout_width = "fill_parent"
        android:layout_height = "fill_parent"
        />
    <LinearLayout
        android:layout_width = "fill_parent"
        android:layout_height = "wrap_content"
        android:orientation = "horizontal"
        >
        <com.shinado.mywhere.view.RadarView
            android:layout_width = "120dip"
            android:layout_height = "120dip"
        />
        <LinearLayout
            android:layout_width = "180dip"
            android:layout_height = "wrap_content"
            android:orientation = "vertical"
            >
            <TextView
              android:layout_width = "fill_parent"
              android:layout_height = "wrap_content"
                />
            <SeekBar
                android:layout_width = "fill_parent"
                android:layout_height = "wrap_content"
                />
        </LinearLayout>
    </LinearLayout>
</RelativeLayout>
```

这里的 WidgetView 也就是我们在第 5 章介绍的控件，在这里就不再进行讲解了，直接拿来使用。

兴趣点 OverlayView 的实现

这一节的内容是本章的重点以及难点，因此我会从重要到次要，再到边边角角来讲解这部分的实现。首先，最重要的当然是兴趣点坐标的计算。

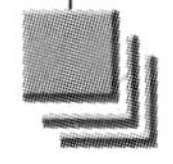

OverlayView 兴趣点坐标的计算

OverlayView 的实现是通过定位数据、方向感应数据、重力感应数据来动态绘制视图，达到增强现实的效果。实现该视图的重点就在于计算出兴趣点在手机界面中的相对应坐标。这个算法封装在 PointManager 中。

首先，我们已知手机定位、重力感应和方向感应的数据，以及兴趣点的经纬度及高度。

那么，以我的位置作为坐标原点，得到兴趣点的相对坐标：

```
/*
 * InterestPoint msg   兴趣点
 * Location userLoc    我的位置
 */
double relativeLon = msg.getLon() - userLoc.getLongitude();
double relativeLat = msg.getLat() - userLoc.getLatitude();
```

接下来计算角度 angle：

```
float angle = getAngle(relativeLon,relativeLat);
//π 转换为 180
angle = (float) (angle * 180/Math.PI);
public float getAngle(double x,double y){
    float ret = (float) Math.abs(Math.atan(x/y));
    //获取象限
    int quadrant = this.getQuadrant(x,y);
    if(quadrant == 4){
        ret = (float) (Math.PI - ret);
    }
    if(quadrant == 3){
        ret += Math.PI;
    }
    if(quadrant == 2){
        ret = (float) (Math.PI * 2 - ret);
    }
    if(ret >= Math.PI * 2){
        ret = 0;
    }
    return ret;
}
```

这个 angle 的值如图 7-7 所示。

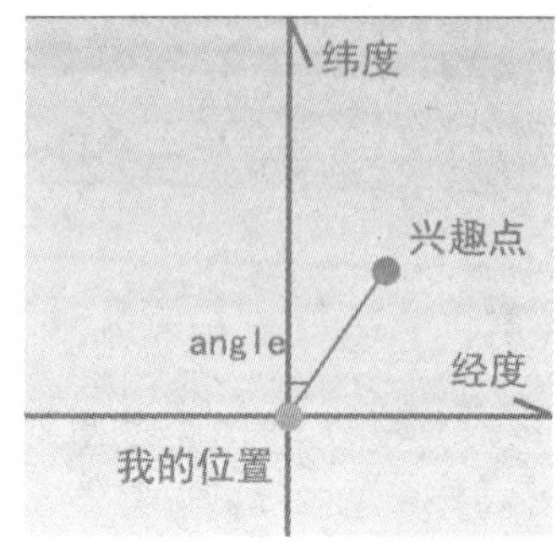

图 7-7　angle 值示意

接下来，获取 x 坐标的值：

```
*
* solid.ori 方向感应器的方向数值
*/
float gap = angle - solid.ori;
/*
* dm.widthPixels  屏幕宽度
* ORI_TO_EDGE     相比值
*/
float moveX = dm.widthPixels/2 * gap/ORI_TO_EDGE;
return moveX;
```

这里的 ORI_TO_EDGE 相比值是指定的一个常量，它指定了偏离多少度时兴趣点会消失在屏幕上，它决定了兴趣点左右晃动的灵敏度，我的设置为 56。在图 7-8 中，大家可以清晰地看出 gap 值的含义(gap＝angle-solid.ori)。

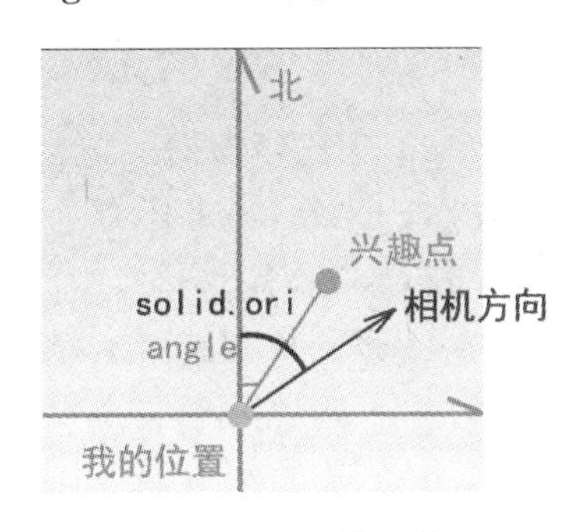

图 7-8　gap 值示意

进行映射，得到公式：

$$\frac{\text{ORI\_TO\_EDGE}}{\text{gqp}}=\frac{\text{dm. widthPixels}}{\text{moveX}}$$

假设 gap 的值为 14，ORI_TO_EDGE 为 56，屏幕的宽度为 480，那么得出偏移量 moveX 为 120。

需要注意的是，moveX 只是偏移量。当 solid. ori 等于 angle 时，此时的兴趣点应该在屏幕的正中央，所以通过这个映射获取的值只是偏移量，要获取真正的坐标还需进行下一步的计算：

```
float moveX = getMoveX(solid,msg,userLoc,dm);
float x = moveX + dm.widthPixels/2;
//Bitmap overlayBm 兴趣点的位图
x = x - overlayBm.getWidth()/2;
```

接下来获取 y 坐标的值。在这之前，我先讲解高度的计算。为了避免远的兴趣点被近的兴趣点覆盖，我设计了一个算法，让远的兴趣点的高度比近的兴趣点要高一些，看起来也比较有层次感一点：

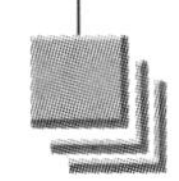

```
//获取与兴趣点的距离
double distance = DistanceUtil.GetDistance(
        mLocation.getLongitude(),mLocation.getLatitude(),
        msg.getLon(),msg.getLat());
//maxDistance 所有兴趣点中最大的距离
double scale = distance/maxDistance;
float height = (float) (5 * scale);
msg.setHeight(height);
```

此外，高度的设置也有另一个作用，让兴趣点可以上下拖动，这一点将在后面讲解。

y坐标偏移量的获取相对简单，也是通过映射获取 moveY 的值：

```
public float getMoveY(Solid solid,InterestPoint msg,DisplayMetrics dm){
    float height = msg.getHeight();
    float acc = solid.acc + height;
    return acc * dm.heightPixels/ACC_TO_EDGE + ADJUST_HEIGHT;
}
```

这里的 ACC_TO_EDGE 设置为 30，这已经超出了重力加速度值 9.8，这是为了保证兴趣点在竖直方向上不会有太大的晃动。此外，这里还有一个 ADJUST_HEIGHT，设置为 50，将兴趣点稍微往上移到屏幕中心位置。

同样，moveY 也是坐标偏移量，需要进行处理获取 y 坐标：

```
loat moveY = getMoveY(solid,msg,dm);
float y = dm.heightPixels/2 - moveY;
//Bitmap overlayBm 兴趣点的位图
y = y - overlayBm.getHeight()/2;
```

这个方法封装在 PointManager 的方法中：

```
public Point getPoint(Solid solid,InterestPoint msg,Bitmap overlayBm,
        Location userLoc,DisplayMetrics dm)
```

OverlayView 的实现

OverlayView 是一个标准的 SurfaceView 架构：

```
public class OverlayView extends SurfaceView implements SurfaceHolder.Callback,
        View.OnTouchListener,Runnable {
```

构造函数也很常规：

```
public OverlayView(Context context,AttributeSet attrs) {
    super(context,attrs);
    this.mContext = context;
    //设置 SurfaceView 相关
    mHolder = this.getHolder();
```

```
        mHolder.addCallback(this);
        mHolder.setFormat(PixelFormat.TRANSPARENT);
        setZOrderOnTop(true);

        //获取屏幕长宽
        dm = new DisplayMetrics();
        ((Activity)context).getWindowManager().
                getDefaultDisplay().getMetrics(dm);

        //初始化
        pointManager = new PointManager();
        mPaint = new Paint();
        mPaint.setAntiAlias(true);
        bitmapUtil = new BitmapUtil(context);
    }
```

在 surfaceCreated 回调函数中，做了两件事：注册感应器和开始绘图的线程。在绘制 OverlayView 时，因为感应器精度的原因，在数值上的一点点变化就会引起图像的偏移，如果直接使用感应器数据作为最终数据，就会产生很强烈的跳跃感。为了解决这个问题，我设计了一个算法，该算法的思想为"一步分十步走"。

在图 7-9 中，直线外的点模拟正常情况下感应器获取的数据，大家可以看到这些点都是在一个方向上的，但是因为在真实的复杂情况下无法形成一个平滑直线，这时候我们就需要进行处理。这个处理的方法就是：以第 1 个数据为起点，第 10 个数据为终点画一条直线（图 7-9 中直线），再在这条直线上平分出 8 个点，此时这些数据都是平滑的。

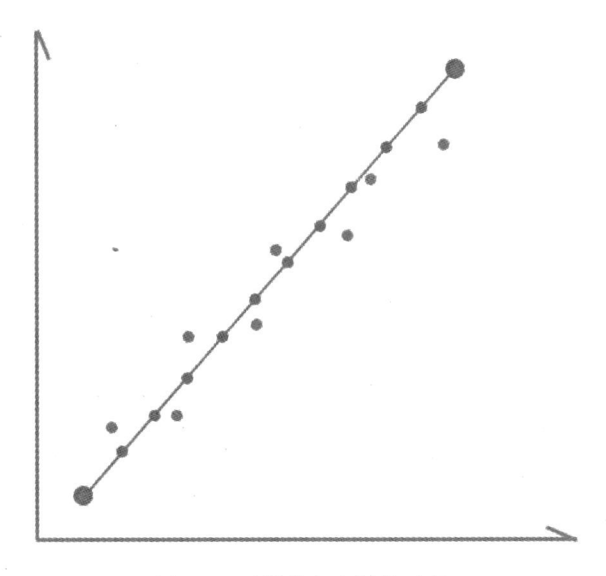

图 7-9 平滑值与实际值示意

在本例中，延长感应器的获取间隔，如果获取间隔为 20 毫秒，那么就变成 200 毫秒。这相当于取第 1 个点与第 10 个点。然后取点，用第 10 个点的数据减去第 1 个点，就是这两点之间的间隔，除以一个常量得到间隔增量，每次绘图的时候都是使用第 1 个的数据加上间隔

增量乘以绘制次数，即可获得所有直线上的点的值，并且是平滑的。

假设感应器获取间隔为 200 毫秒，绘图间隔为 20 毫秒，那么就应该将这段数据平分为 10 段。

好了，讲解完毕之后我们来看代码。首先是感应器数据的获取：

```
public class WSensorManager {

    private SensorManager sm;
    private SensorChangeListener sensorListner;

    private float mOri;
    private float mAcc;

    private boolean mFlag = true;

    public static final float SLEEP_TIME = 200;

    public void register(Context context) {
        sm = (SensorManager)context.getSystemService(Context.SENSOR_SERVICE);
        start();
        new Thread(){
            public void run(){
                while(mFlag){
                    //监听器
                    if(sensorListner != null){
                        sensorListner.onChange(mOri,mAcc);
                    }
                    //感应器数据获取间隔时间
                    try {
                        Thread.sleep((int)SLEEP_TIME);
                    } catch (InterruptedException e) {
                        e.printStackTrace();
                    }
                }
            }
        }.start();
    }

    public void unregister()
    {
        stop();
        mFlag = false;
    }

    SensorEventListener accListener = new SensorEventListener()
    {
        @Override
        public void onAccuracyChanged(Sensor sensor, int accuracy) {
        }
```

```
        @Override
        public void onSensorChanged(SensorEvent event) {
            mAcc = event.values[2];
        }
    };
    SensorEventListener oriListener = new SensorEventListener()
    {
        @Override
        public void onAccuracyChanged(Sensor sensor, int accuracy) {
        }

        @Override
        public void onSensorChanged(SensorEvent event) {
            mOri = event.values[0];
        }
    };

    public void setSensorListner(SensorChangeListener sensorListner) {
        this.sensorListner = sensorListner;
    }

    private void stop(){
        sm.unregisterListener(accListener);
        sm.unregisterListener(oriListener);
    }
    private void start(){
        sm.registerListener(accListener,
                sm.getDefaultSensor(Sensor.TYPE_ACCELEROMETER),
                    SensorManager.SENSOR_DELAY_UI);

        sm.registerListener(oriListener,
            sm.getDefaultSensor(Sensor.TYPE_ORIENTATION),
                    SensorManager.SENSOR_DELAY_NORMAL);
    }
    public interface SensorChangeListener{
        public void onChange(float ori, float acc);
    }
}
```

接下来，在 surfaceCreated 回调函数中计算间隔增量：

```
@Override
public void surfaceCreated(SurfaceHolder holder) {
    //注册感应器监听
    sm = new WSensorManager();
    sm.register(mContext);
    sm.setSensorListner(new SensorChangeListener(){
        @Override
        public void onChange(float ori, float acc, float skew) {
            /*
             * 获取平稳值
             */
```

```
                Solid solid = new Solid();
                solid.ori = ori;
                solid.acc = acc;
                if(solidList.size() == 2){
                    solidList.remove(0);
                }
                solidList.add(solid);
                if(solidList.size() == 2){
                        //间隔增量
                        mAccAdd = (solidList.get(1).acc - solidList.get(0).acc)
                              /(WSensorManager.SLEEP_TIME/SCALE_RATIO);
                    mOriAdd = (solidList.get(1).ori - solidList.get(0).ori)
                              /(WSensorManager.SLEEP_TIME/SCALE_RATIO);
                  }
                scale = 1;
            }
        });
        mIsRun = true;
        new Thread(this).start();
    }
```

最后，在线程中计算直线上的点并绘图：

```
    @Override
    public void run() {
        while(mIsRun){
            if(mIsToDraw){
                if(solidList.size() == 2){
                    mSolid.acc = solidList.get(0).acc + mAccAdd * scale;
                    mSolid.ori = solidList.get(0).ori + mOriAdd * scale;
                    scale++;
                    draw();
                }
            }
        }
    }
```

我们先看绘图函数，再回过头来讲兴趣点的加载、位图的绘制以及缩放大小的设置：

```
    private void draw(){
        Canvas canvas = mHolder.lockCanvas();
        try {
            canvas.drawColor(Color.TRANSPARENT, Mode.CLEAR);
        } catch (Exception e) {
            return;
        }
        mIsLoaded = false;
        for(int i = 0; i < mMsgs.size(); i++){
            InterestPoint msg = mMsgs.get(i);
```

```
        //获取位图
        Bitmap overlayBitmap = msg.getImageBm();
        //获取绘图坐标
        Point point = pointManager.getPoint(mSolid,msg,
                overlayBitmap,mLocation,dm);
        //获取缩放大小
        float mx = msg.getMx();
        if(Math.abs(mx - ( - 1))< 0.1f){
            continue;
        }
        /*
         * mIndex 用户触摸的兴趣点序号
         * 如果用户触摸该兴趣点,将它缩小
         */
        if(mIndex == i){
            mx - = 0.1f;
        }

        //保存位图的位置
        Overlay overlay = overlays.get(i);
        Rect rect = overlay.getRect();
        rect.left = point.x;
        rect.top = point.y;
        rect.right = point.x + (int) (overlayBitmap.getWidth() * mx);
        rect.bottom = point.y + (int) (overlayBitmap.getHeight() * mx);
        overlay.setRect(rect);

        //如果超出屏幕就不绘图
        if(rect.right < 0 || rect.left > dm.widthPixels ||
          rect.top > dm.heightPixels || rect.bottom < 0){
            continue;
        }

        //缩放
        mMatrix.reset();
        mMatrix.setTranslate(point.x,point.y);
        mMatrix.preScale(mx,mx);

        canvas.drawBitmap(overlayBitmap,mMatrix,mPaint);

    }
    mIsLoaded = true;
    mHolder.unlockCanvasAndPost(canvas);
}
```

这里的位置信息 mLocation 由 Activity 注入。每当有位置更新时,重新设置兴趣点:

```
public void setLocation(Location location){
    mLocation = location;
```

```
        addMsg(mMsgs);
        radarView.setLocation(location);
        radarView.addDialogMsgs(mMsgs);
    }
```

兴趣点的加载总共做了 4 件事：①清除原有数据；②与雷达视图同步；③创建位图；④计算最远兴趣点与我的距离。由于雷达和兴趣点视图是一一对应的，所以雷达视图的所有数据应该是与兴趣点视图同步的：

```
    public void addMsg(final ArrayList<InterestPoint> msgs) {
        mIsToDraw = false;
        clearData();
        //更新 RadarView
        radarView.addDialogMsgs(msgs);

        new Thread(){
            @Override
            public void run(){
                //创建位图
                createBitmap(msgs);
                //计算最远兴趣点离我的距离
                getMaxDistance();
                mIsToDraw = true;
            }
        }.start();
    }
```

首先来看计算最远兴趣点的距离：

```
    private void getMaxDistance(){
        for(int i = 0; i < mMsgs.size(); i++){
            InterestPoint msg = mMsgs.get(i);
            double distance = msg.getDistance();
            if(maxDistance < distance){
                maxDistance = distance;
            }
        }
        refreshMatrix();
    }
```

refreshMatrix()这个函数用于计算缩放值以及计算高度：

```
    private void refreshMatrix(){
        for(int i = 0; i < mMsgs.size(); i++){
            InterestPoint msg = mMsgs.get(i);
            Location msgLocation = new Location("");
            msgLocation.setLatitude(msg.getLat());
            msgLocation.setLongitude(msg.getLon());
```

```
            float mx = getMatrix(msgLocation,mLocation,maxDistance);
            msg.setMx(mx);

            //获取与兴趣点的距离
            double distance = msg.getDistance();
            //maxDistance 所有兴趣点中最大的距离
            double scale = distance/maxDistance;
            float height = (float) (5 * scale);
            msg.setHeight(height);
        }
    }
    private static final float MIN_MATRIX = 0.35F;
    private static final float DEFAULT_DISTANCE_MAX = 1000F;
    private float getMatrix(Location msg,Location user,double maxDistance) {
        double distance = DistanceUtil.getDistance(
                msg.getLongitude(),msg.getLatitude(),
                user.getLongitude(),user.getLatitude());
        if(Math.abs(maxDistance) > 0.1f){
            if(distance > maxDistance){
                return -1;
            }
        }else{
            maxDistance = DEFAULT_DISTANCE_MAX;
        }
        float mx = (float) (maxDistance/distance)/2;
        if(mx > 1){
            mx = 1;
        }else if(mx < MIN_MATRIX){
            mx = MIN_MATRIX;
        }
        return mx;
    }
```

接下来来讲创建位图的过程，它包括：①计算距离；②通过地址以及兴趣点名称生成Bitmap；③设置 overlays 对象；④保存兴趣点对象。

```
    private void createBitmap(ArrayList<InterestPoint> msgs){
        for(int i = 0; i < msgs.size(); i++){
            //获取兴趣点
            InterestPoint msg = msgs.get(i);
            //计算距离
            double distance = DistanceUtil.getDistance(
                    mLocation.getLongitude(),mLocation.getLatitude(),
                    msg.getLon(),msg.getLat());
            msg.setDistance(distance);
            //生成 Bitmap 位图
            Bitmap result = bitmapUtil.drawTextInDialog(msg);
            msg.setImageBm(result);
```

```
            //设置 overlays 对象
            Overlay overlay = new Overlay();
            overlay.setMsg(msg);
            overlay.setBitmap(result);
            overlay.setRect(new Rect());
            overlays.add(overlay);
             //保存兴趣点对象
            mMsgs.add(msg);
        }
    }
```

overlays这个对象保存了兴趣点的位图以及位置信息，它的功能有两个：①判断用户是否触摸该兴趣点；②拍照时根据overlays对象绘制兴趣点。需要注意在draw()函数执行的过程中，overlays对象会变化，所以在获取overlays对象时，需要等待draw()函数执行完毕后再获取：

```
public ArrayList<Overlay> getOverlays() {
    while(mIsLoaded == false || overlays == null);
    return overlays;
}
```

就第一点来说，我们来看触摸事件。当ACTION_DOWN时，判断是否触摸到了某一个兴趣点，如果有的话记录在mIndex中：

```
case MotionEvent.ACTION_DOWN:
    ArrayList<Overlay> overlays = getOverlays();
    for(int i = overlays.size() - 1; i >= 0; i--){
        Rect rect = overlays.get(i).getRect();
        if(rect.contains(x,y)){
            mIndex = i;
            return true;
        }
    }
    break;
```

如果触摸了某个兴趣点，则通过改变height值来拖动兴趣点：

```
case MotionEvent.ACTION_MOVE:
    if(mIndex >= 0){
        InterestPoint msg = mMsgs.get(mIndex);
        float height = pointManager.returnHeight(mSolid,y,dm);
        msg.setHeight(height);
        mTouchTimes++;
    }
    break;
```

获取height值完全是获取y坐标的逆过程：

```
public float returnHeight(Solid solid, int y, DisplayMetrics dm){
    float moveY = dm.heightPixels/2 - y - ADJUST_HEIGHT;
    return moveY/(dm.heightPixels/ACC_TO_EDGE) - solid.acc;
}
```

ACTION_UP 时，进行判断并归位：

```
case MotionEvent.ACTION_UP:
    if(overlayTouchListener != null && mIndex >= 0){
        //移动次数少于5,表示是单击而不是拖动
        if(mTouchTimes <= 5){
            overlayTouchListener.onTouch(getOverlays().get(mIndex));
        }
    }
    mIndex = -1;
    mTouchTimes = 0;
    break;
}
```

对于第二点拍照，在 Activity 中定义拍照函数：

```
mCameraView.setPicCallback(new PictureCallback(){
    @Override
    public void onPictureTaken(final byte[] data, Camera camera) {
        //等待对话框
        mpd.setMessage("正在保存图片");
        mpd.show();

        new Thread(){
            public void run(){
                //获取相片位图
                BitmapFactory.Options options = new BitmapFactory.Options();
                options.inSampleSize = 1;
                Bitmap photo = BitmapFactory.decodeByteArray(
                        data,0,data.length,options);

                //在照片上面绘制兴趣点
                BitmapUtil bmUtil = new BitmapUtil(ARActivity.this);
                Bitmap result = bmUtil.drawTagOnPhoto(
                        photo,mOverlayView.getOverlays());

                //保存图片
                String time = TimeUtil.getTimeInSec();
                bmUtil.saveBitmap(PIC_PATH,time,result);

                mOverlayView.stopDrawing();
                mpd.dismiss();
                mHandler.sendEmptyMessage(1);
            }
```

```
            }.start();
        }
    });
```

还有最后一点，雷达缩放功能的实现，这里使用了函数，通过设置最远值以及重新计算缩放的比例来实现：

```
public void setMaxDistance(double distance) {
    this.maxDistance = distance;
    radarView.setMaxDistance(distance);
    refreshMatrix();
}
```

此外，还有一个函数 refreshBitmapInOverlay()，这个函数用于在使用雷达进行缩放时，重新更新位图的大小：

```
private static final int MIN_WIDTH = 10;
public void refreshBitmapInOverlay(){
    for(int i = 0; i < mMsgs.size(); i++){
        InterestPoint msg = mMsgs.get(i);
        Bitmap bm = msg.getImageBm();
        Location msgLocation = new Location("");
        msgLocation.setLatitude(msg.getLat());
        msgLocation.setLongitude(msg.getLon());
        float mx = getMatrix(msgLocation, mLocation, maxDistance);
        if(mx < 0){
            bm = null;
        }else{
            if(bm != null){
                mMatrix.reset();
                mMatrix.postScale(mx, mx);
                if(bm.getWidth() <= MIN_WIDTH || bm.getHeight() <= MIN_WIDTH){
                    bm = Bitmap.createBitmap(bm, 0, 0,
                            MIN_WIDTH, MIN_WIDTH, mMatrix, false);
                }else{
                    bm = Bitmap.createBitmap(bm, 0, 0,
                            bm.getWidth(), bm.getHeight(), mMatrix, false);
                }
            }
        }
        Overlay overlay = overlays.get(i);
        overlay.setBitmap(bm);
        overlay.setRect(new Rect());
    }
}
```

雷达的实现

雷达的实现较为简单，也是一个 SurfaceView，这里使用了两个位图，如图 7-10。

图 7-10 雷达使用的图片资源

在应用中，它看起来如图 7-11。

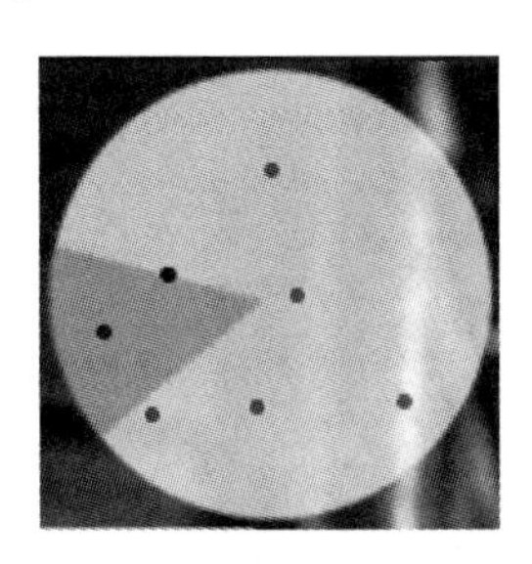

图 7-11 雷达视图最终效果

不知道大家还记不记得在讲兴趣点的坐标计算时，ORI_TO_EDGE 的值为 56？这里的 compass_fov 的阴影区域夹角正是 56°。

与兴趣点视图相同，在 surfaceCreated 回调函数中添加监听器：

```
@Override
public void surfaceCreated(SurfaceHolder holder) {
    createBitmap();
    sm = new WSensorManager();
    sm.setSensorListner(new SensorChangeListener(){
        @Override
        public void onChange(float ori,float acc) {
            if(!mIsDestroyed){
                try {
                    draw(ori);
                } catch (Exception e) {
                    e.printStackTrace();
                }
            }
        }
    });
    sm.register(mContext);
    mIsDestroyed = false;
}
```

在 createBitmap()函数中，初始化位图。为了保证雷达视图可以由配置文件设定而变化，因此在这个函数中设定了位图的缩放：

```
private void createBitmap(){
    mPlainBm = BitmapFactory.decodeResource(mContext.getResources(),
        R.drawable.compass_plain);
    mFovBm = BitmapFactory.decodeResource(mContext.getResources(),
        R.drawable.compass_fov);
  /*
    * 在布局文件中限定了大小
    */
    float width = this.getWidth();
    float height = this.getHeight();
    //比实际半径稍微小一点
    max = width/2 - width/2 * 0.1f;

    //按比例缩放位图
    Matrix matrix = new Matrix();
    int bmWidth = mPlainBm.getWidth();
    int bmHeight = mPlainBm.getHeight();
    matrix.postScale(width/bmWidth, height/bmHeight);
    mPlainBm = Bitmap.createBitmap(mPlainBm,0,0,bmWidth,bmHeight,
            matrix,false);
    mFovBm = Bitmap.createBitmap(mFovBm,0,0,bmWidth,bmHeight,
            matrix,false);
}
```

在绘图函数中绘制罗盘、扫射区域以及兴趣点：

```
private void draw(float ori){
    Canvas canvas = mHolder.lockCanvas();
    canvas.drawColor(Color.TRANSPARENT,Mode.CLEAR);
    //旋转雷达蓝色扫射区域
    matrix.reset();
    matrix.preRotate(ori,
        (float)mFovBm.getWidth()/2,(float)mFovBm.getHeight()/2);
    //mPlainBm compass_plain.png
    canvas.drawBitmap(mPlainBm,0,0,null);
    //mFovBm compass_fov.png
    canvas.drawBitmap(mFovBm,matrix,mPaint);
    //绘制兴趣点(小黑点)
    drawDots(canvas);

    mHolder.unlockCanvasAndPost(canvas);
}
```

兴趣点的绘制要稍微复杂一些，但思路也很清晰：

(1) 计算兴趣点与我的位置的角度；

(2) 根据比例确定兴趣点在雷达上的半径。

比例计算公式为：

$$\frac{\text{雷达半径}}{\text{最远距离}}=\frac{\text{兴趣点半径}}{\text{兴趣点距离}}$$

(3) 根据三角函数计算相对于我的坐标。

(4) 转换成雷达上的坐标。

```
private void drawDots(Canvas canvas){
    for(int i = 0; i < mMsgs.size(); i++){
        InterestPoint msg = mMsgs.get(i);
        double lat = msg.getLat();
        double lon = msg.getLon();
        //(1)计算角度
        float angle = pointManager.getAngle(
                lon - mLocation.getLongitude(),
                lat - mLocation.getLatitude());

        //计算实际距离
        double distance = DistanceUtil.getDistance(
                lon,lat,mLocation.getLongitude(),
                  mLocation.getLatitude());
        /*
         * max 雷达半径
         * (2)根据比例计算兴趣点在雷达上的半径
         */
        distance = distance * max/maxDistance;
        if(distance < MIN){
            distance = MIN;
        }
        if(distance > max){
            continue;
        }
        //(3)获得相对于我的位置的坐标
        Point point = pointManager.getPoint(distance,angle);
        //(4)获得绘图坐标
        float x = point.x + mPlainBm.getWidth()/2;
        float y = mPlainBm.getHeight()/2 - point.y;
        canvas.drawCircle(x,y,DOT_SIZE,mPaint);
    }
}
```

此外，雷达视图还有一个功能，就是通过拖动滑动条来调整，调整时，兴趣点会跟着缩放，如图 7-12。

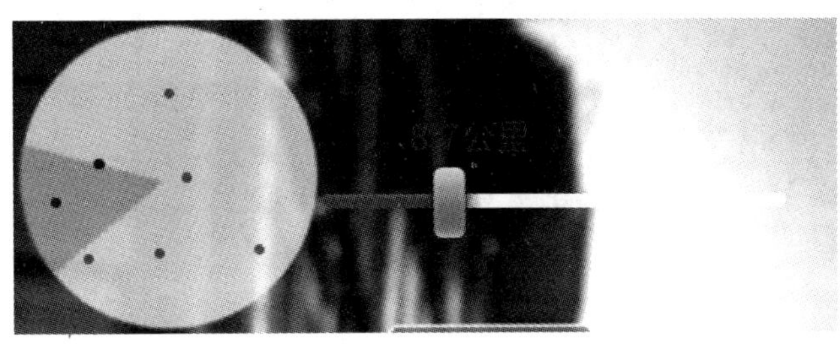

图 7-12　滑块调整兴趣点缩放

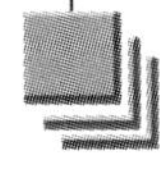

实现的代码为：

```
distanceSelector = (SeekBar) findViewById(R.id.distanceSelector);
distanceSelector.setOnSeekBarChangeListener(new OnSeekBarChangeListener(){

    @Override
    public void onProgressChanged(SeekBar arg0, int progress, boolean arg2) {
        double maxDistance = (float)progress/PROGRESS_MAX * (mMaxDistance * 2);
        if(maxDistance <= 1){
            maxDistance = 1;
        }
        mOverlayView.setMaxDistance(maxDistance);
        String text = formateDistance(maxDistance, progress);
        distanceTv.setText(text);
    }

    @Override
    public void onStartTrackingTouch(SeekBar arg0) {
    }
    @Override
    public void onStopTrackingTouch(SeekBar arg0) {
        mOverlayView.refreshBitmapInOverlay();
    }
});
```

现实视图 ARActivity 的实现

1. 定位的实现

```
public class MyLocationProvider implements LocationListener{

    LocationManager mLocationManager;
    Location mLocation = null;
    private static final String DEFAULT_LOCATION_PROVIDER =
        LocationManager.NETWORK_PROVIDER;
    String strLocationProvider = DEFAULT_LOCATION_PROVIDER;

    private OnLocationChangedListener listener;

    public MyLocationProvider(Context context)
    {
         mLocationManager = (LocationManager)
            context.getSystemService(Context.LOCATION_SERVICE);
         String provider = getProvider();
        if(provider != null){
            strLocationProvider = provider;
        }
        mLocationManager.requestLocationUpdates(
```

```
            strLocationProvider,2000,50,this);
    }

    private String getProvider(){
        //设置定位策略
        Criteria mCriteria = new Criteria();
        mCriteria.setAccuracy(Criteria.ACCURACY_FINE);
        mCriteria.setAltitudeRequired(false);
        mCriteria.setBearingRequired(false);
        mCriteria.setCostAllowed(true);
        mCriteria.setPowerRequirement(Criteria.POWER_MEDIUM);
        return mLocationManager.getBestProvider(mCriteria,true);
    }
    public void initLocation(){
    long startTime = System.currentTimeMillis();
    while(mLocation == null){
            //可能要多次才能获得上一次定位
            mLocation = mLocationManager.getLastKnownLocation(strLocationProvider);
            try {
                Thread.sleep(1000);
            } catch (InterruptedException e) {
                e.printStackTrace();
            }
            if(System.currentTimeMillis() - startTime > 5000){
                strLocationProvider = DEFAULT_LOCATION_PROVIDER;
            }
            if(System.currentTimeMillis() - startTime > 10000){
                break;
            }
        }
    }

    @Override
    public void onLocationChanged(Location location) {
        if(listener != null){
            listener.onLocationChanged(location);
        }
    }

    @Override
    public void onProviderDisabled(String arg0) {
    }

    @Override
    public void onProviderEnabled(String arg0) {
    }

    @Override
    public void onStatusChanged(String arg0,int arg1,Bundle arg2) {
    }
```

```
    public Location getMyLocation()
    {
        return mLocation;
    }

    public void setListener(OnLocationChangedListener listener) {
        this.listener = listener;
    }

    public interface OnLocationChangedListener{
        public void onLocationChanged(Location location);
    };
}
```

2. 搜索的实现

搜索小窗口从控件中打开，如图 7-13 所示。

图 7-13　搜索窗口

这个搜索小对话框其实是一个 Activity，只不过它的主题进行了特殊的设置：

```
<activity android:name = "com.shinado.mywhere.SearchActivity"
    android:theme = "@android:style/Theme.Translucent.NoTitleBar"/>
```

它的布局很简单，在根布局中设定长宽即可：

```
<?xml version = "1.0" encoding = "utf - 8"?>
<com.shinado.mywhere.view.CancelOnTouchLayout
    android:layout_height = "45dp"
    android:layout_marginTop = "20dp"
    android:layout_width = "300dp" >

    <EditText
        android:layout_width = "fill_parent"
        android:layout_height = "wrap_content"
        />
    <ImageButton
        android:onClick = "search"/>
</com.shinado.mywhere.view.CancelOnTouchLayout>
```

CancelOnTouchLayout 这个类继承于 RelativeLayout，它的功能顾名思义：

```
public class CancelOnTouchLayout extends RelativeLayout{
```

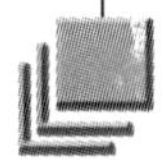

```
    public CancelOnTouchLayout(final Context context,AttributeSet attrs) {
        super(context,attrs);
        this.setOnClickListener(new OnClickListener() {
            @Override
            public void onClick(View v) {
                ((Activity)context).finish();
            }
        });
    }
}
```

在控件事件中启动这个 Activity：

```
startActivityForResult(
        new Intent(ARActivity.this,SearchActivity.class),REQUEST_SEARCH);
```

在回调函数中添加搜索功能：

```
@Override
public void onActivityResult(int requestCode,int resultCode,Intent intent){
    switch(requestCode){
    case REQUEST_SEARCH:
        if(resultCode == RESULT_DONE){
            String key = intent.getStringExtra("key");
            Location location = mLp.getMyLocation();
            GeoPoint pt = new GeoPoint((int) (location.getLatitude() * 1E6),
                    (int) (location.getLongitude() * 1E6));
            int result = mkSearch.poiSearchNearBy(key,pt,10000);
        }
        break;
    }
}
```

搜索功能 API 由百度提供：

```
private BMapManager mBMapMan = null;
private MKSearch mkSearch = null;
//搜索功能的初始化
private void initSearch(){
    mkSearch = new MKSearch();
    mBMapMan = new BMapManager(this);
    mBMapMan.init("E4C1F14AFE60CE7C0BA562130B483E5B98FDD537",null);
    mkSearch.init(mBMapMan,mkSearchListener);
}
MKSearchListener mkSearchListener = new MKSearchListener() {
    //省略其他 override 的函数
    @Override
    public void onGetPoiResult(MKPoiResult arg0,int arg1,int arg2) {
        //首先判断是否搜索到结果
```

```
            if(arg2 != 0 || arg0 == null)
            {
                Toast.makeText(ARActivity.this,"没有找到结果!",
                        Toast.LENGTH_SHORT).show();
                return;
            }
            //添加兴趣点
            if(arg0.getCurrentNumPois() > 0)
            {
                ArrayList<MKPoiInfo> list = arg0.getAllPoi();
                if(list.size() == 0){
                    return;
                }
                mMsgs.clear();
                for(MKPoiInfo info:list){
                    mMsgs.add(new InterestPoint(
                        info.uid,info.name,info.address,info.pt));
                }
                mOverlayView.addMsg(mMsgs);
                getMaxDistance(mLp.getMyLocation());
            }
        }
    };
```

第八章　电子菜单系统

体验 Android 平板的大气设计风范

清单

Demo 代码：

```
\demo\Demo_YiRstr
\demo\Demo_YiRstr_2
```

实例代码：

```
\source_codes\YiRstr
```

Target SDK：

```
Android 3.0
```

第一节　产品介绍

本实例来源于实际项目，由于篇幅有限，我只把里面的核心功能提取出来，去掉了网络传输功能。真正意义上的平板点餐系统需要使用网络，但是在这里我把网络的模块给剔除了，以便于讲解。读者如果有兴趣的话可以继续把该项目完善，成为一个真正的点餐系统。

功能分析

日程应用有以下几个功能：

- 查看菜单；
- 菜单分类；
- 购物车；
- 查看订单；
- 音乐播放；
- 呼叫服务。

界面设计

借助平板屏幕大的特点，有更多的空间来摆放控件，在设计的时候体现出大气的特点，如图 8-1 所示。

图 8-1　系统界面

本实例采用单 Activity 的方式，所有的视图都在一个 Activity 上完成。

fragment

fragment 是 Android 3.0 以后开始的全新控件。说直观一点，Android 3.0 中的设置界面就是 fragment 的一个实例，如图 8-2 所示。

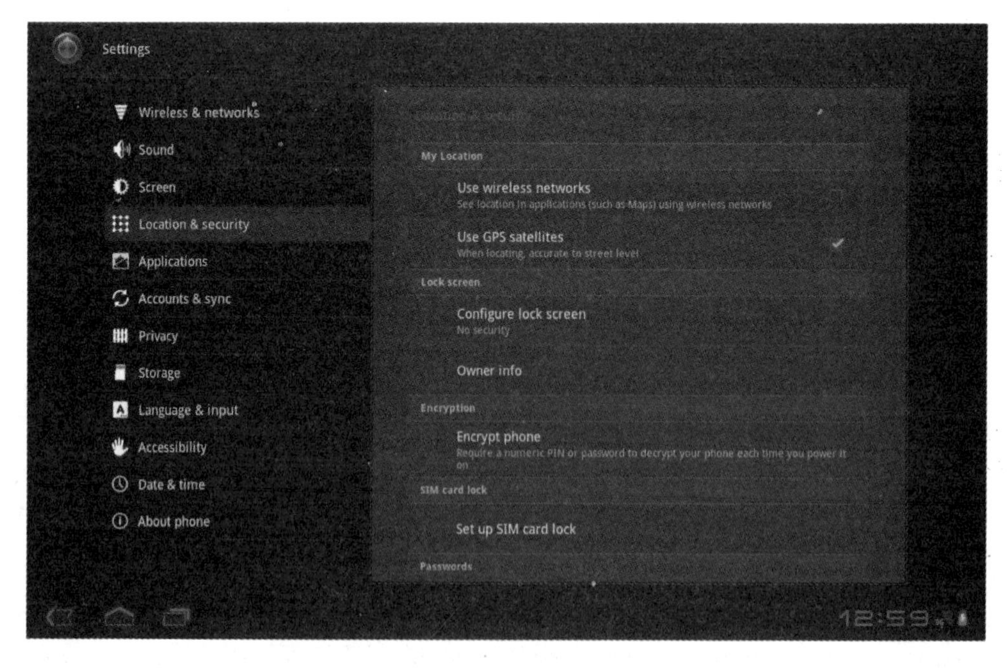

图 8-2　设置界面——fragment 的一个应用

新建一个 Master/Detail Flow

以 Android 3.0 作为 SDK，这时候就可以选择 Holo Dark 或者 Holo Light 的 Theme 了，如图 8-3 所示。这个特性是从 Android 3.0 后才开始的。

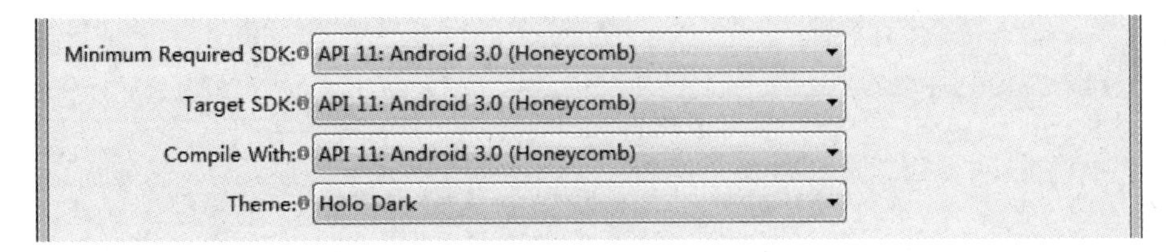

图 8-3　选择 SDK 以及 Theme

接下来选择 Master/Detail Flow，也就是包含 fragment 的 Activity，如图 8-4 所示。

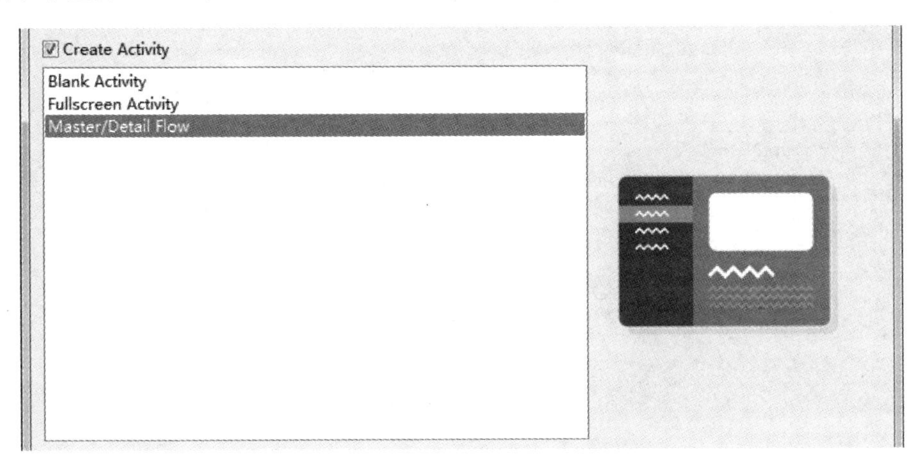

图 8-4　选择 Master/Detail Flow

最后一步，修改 Object，如图 8-5 所示，这个无关紧要。

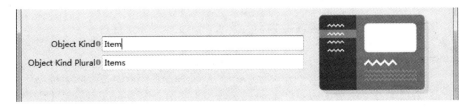

图 8-5　修改 Object

创建成功的工程会比较复杂，一共有 5 个类和 4 个 layout，如图 8-6 所示。

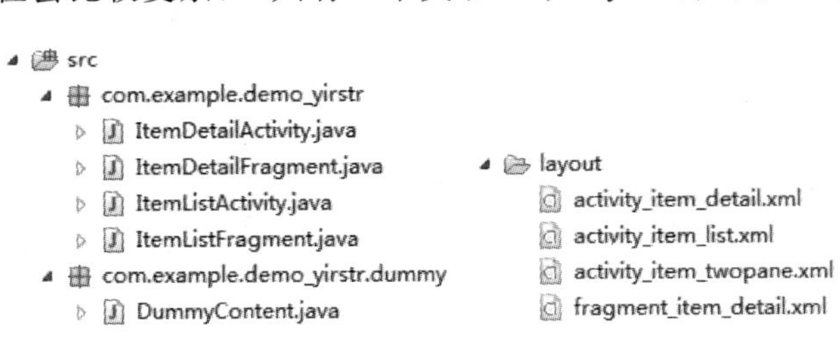

图 8-6　创建的工程初始文件

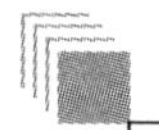

代码分析

我们先从程序的入口来分析这些代码；从 AndroidManifest. xml 文件中得知程序的入口为 ItemListActivity：

```
public class ItemListActivity extends FragmentActivity implements
    ItemListFragment.Callbacks {

    /**
     * activity是否是"两个面板"模式(也就是运行在平板上)
     */
    private boolean mTwoPane;

    @Override
    protected void onCreate(Bundle savedInstanceState) {
        super.onCreate(savedInstanceState);
        setContentView(R.layout.activity_item_list);

        if (findViewById(R.id.item_detail_container) != null) {
            //运行在平板设备上
            mTwoPane = true;
            //设置 fragment 的属性
            ((ItemListFragment) getSupportFragmentManager().findFragmentById(
                    R.id.item_list)).setActivateOnItemClick(true);
        }
    }

    @Override
    public void onItemSelected(String id) {
        if (mTwoPane) {
            //显示相应内容
            Bundle arguments = new Bundle();
            arguments.putString(ItemDetailFragment.ARG_ITEM_ID, id);
            ItemDetailFragment fragment = new ItemDetailFragment();
            fragment.setArguments(arguments);
            getSupportFragmentManager().beginTransaction()
                    .replace(R.id.item_detail_container, fragment).commit();

        } else {
            //在不是平板设备上,开启另一个 activity
            Intent detailIntent = new Intent(this, ItemDetailActivity.class);
            detailIntent.putExtra(ItemDetailFragment.ARG_ITEM_ID, id);
            startActivity(detailIntent);
        }
    }
}
```

我一开始看这段代码时也很迷茫，迷茫是必须的，因为有好几个知识点没有涉及到。我们来慢慢分析这些代码。

先来看这个 Activity 对应的 layout：

```
<fragment xmlns:android = "http://schemas.android.com/apk/res/android"
    xmlns:tools = "http://schemas.android.com/tools"
    android:id = "@+ id/item_list"
    android:name = "com.example.demo_yirstr.ItemListFragment"
    android:layout_width = "match_parent"
    android:layout_height = "match_parent"
    android:layout_marginLeft = "16dp"
    android:layout_marginRight = "16dp"
    tools:context = ".ItemListActivity"
    tools:layout = "@android:layout/list_content"/>
```

可是为什么没有看到 item_detail_container 这个 id？明明在代码中有这么一句：

```
if (findViewById(R.id.item_detail_container) != null) {
```

这时候就要看 values 文件夹了，如图 8-7 所示。

values-large
refs.xml
values-sw600dp
refs.xml

图 8-7　values 文件夹

refs.xml 的代码如下：

```
<resources>
    <item name = "activity_item_list" type = "layout">
        @layout/activity_item_twopane</item>
</resources>
```

读者可能看出点端详了。这两个文件夹保存的配置了程序在大屏下的参数，在这里就表现为把 activity_item_list 这个 layout 替换为 activity_item_twopane。也就是说，这个程序在平板上和手机上的表现是不一样的，如图 8-8 和图 8-9 所示。

图 8-8　在平板上的视图

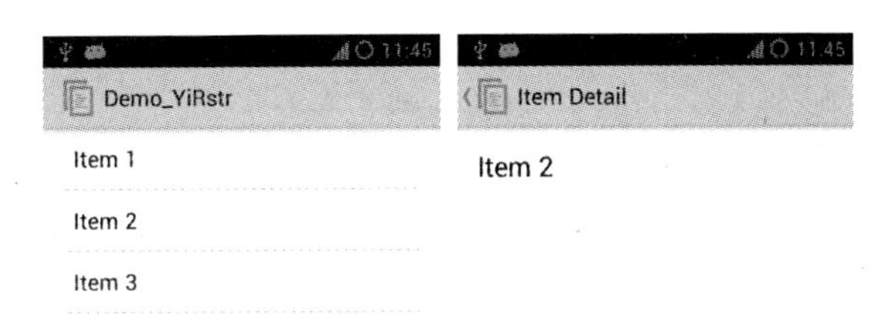

图 8-9　在手机上的视图

我们只看与平板相关的代码，所以 activity_item_list，activity_item_detail 和 ItemDetailActivity 我们都可以不管了。

在平板上，这个 activity 的 layout 其实是 activity_item_twopane：

```
<LinearLayout xmlns:android = "http://schemas.android.com/apk/res/android"
    xmlns:tools = "http://schemas.android.com/tools"
    android:layout_width = "match_parent"
    android:layout_height = "match_parent"
    android:layout_marginLeft = "16dp"
    android:layout_marginRight = "16dp"
    android:baselineAligned = "false"
    android:divider = "?android:attr/dividerHorizontal"
    android:orientation = "horizontal"
    android:showDividers = "middle"
    tools:context = ".ItemListActivity" >

    <fragment
        android:id = "@+id/item_list"
        android:name = "com.example.demo_yirstr.ItemListFragment"
        android:layout_width = "0dp"
        android:layout_height = "match_parent"
        android:layout_weight = "1"
        tools:layout = "@android:layout/list_content"/>

    <FrameLayout
        android:id = "@+id/item_detail_container"
        android:layout_width = "0dp"
        android:layout_height = "match_parent"
        android:layout_weight = "3"/>

</LinearLayout>
```

先看 fragment 的这个属性：

```
android:name = "com.example.demo_yirstr.ItemListFragment"
```

这个属性指定了 fragment 的 Fragment 为 ItemListFragment，这个 Fragment 主要做了两件事：①设置 layout；②实现单击事件（在这里又抽象了一个事件，让 activity 去实现）。当然，除了这两件事之外，为了保证程序的完整性，还重写了 onViewCreated 等方法，用于应对程序被系统杀死的情况，为了让代码清晰一些，我就只列出设置 layout 与实现单击事件的代码：

```
public class ItemListFragment extends ListFragment {
    /**
     * 让外部 activity 实现的 callback,
     * 用来把数据传递给外部 activity
     */
```

```
public interface Callbacks {
    public void onItemSelected(String id);
}
private Callbacks mCallbacks = sDummyCallbacks;

/**
 * 默认的 callback
 */
private static Callbacks sDummyCallbacks = new Callbacks() {
    @Override
    public void onItemSelected(String id) {
    }
};

public ItemListFragment() {
}

@Override
public void onCreate(Bundle savedInstanceState) {
    super.onCreate(savedInstanceState);

    //设置一个最简单的 layout
    setListAdapter(new ArrayAdapter<DummyContent.DummyItem>(getActivity(),
            android.R.layout.simple_list_item_activated_1,
            android.R.id.text1, DummyContent.ITEMS));
}

@Override
public void onAttach(Activity activity) {
    super.onAttach(activity);

    //attach 的 activity 必须实现 callback
    if (!(activity instanceof Callbacks)) {
        throw new IllegalStateException(
                "Activity must implement fragment's callbacks.");
    }

    mCallbacks = (Callbacks) activity;
}

@Override
public void onDetach() {
    super.onDetach();
    //设置一个默认的 callback
    mCallbacks = sDummyCallbacks;
}

@Override
public void onListItemClick(ListView listView, View view, int position,
        long id) {
```

```
        super.onListItemClick(listView,view,position,id);
        //传递给外部 activity
        mCallbacks.onItemSelected(DummyContent.ITEMS.get(position).id);
    }
```

记不记得在 ItemListActivity 中实现了 ItemListFragment.Callbacks 接口并重写它的方法：

```
@Override
public void onItemSelected(String id) {
    if (mTwoPane) {
        //显示相应内容
        Bundle arguments = new Bundle();
        arguments.putString(ItemDetailFragment.ARG_ITEM_ID,id);
        ItemDetailFragment fragment = new ItemDetailFragment();
        fragment.setArguments(arguments);
        //用这个 fragment 替换 item_detail_container
        getSupportFragmentManager().beginTransaction()
                .replace(R.id.item_detail_container,fragment).commit();
    } else {
        //在不是平板设备上,开启另一个 activity
    }
}
```

这里又有另一个 fragment 了。它是 ItemDetailFragment：

```
public class ItemDetailFragment extends Fragment {

    public static final String ARG_ITEM_ID = "item_id";

    /**
     *数据
     */
    private DummyContent.DummyItem mItem;

    @Override
    public void onCreate(Bundle savedInstanceState) {
        super.onCreate(savedInstanceState);
        if (getArguments().containsKey(ARG_ITEM_ID)) {
            //获取数据
            mItem = DummyContent.ITEM_MAP.get(getArguments().getString(
                    ARG_ITEM_ID));
        }
    }

    @Override
    public View onCreateView(LayoutInflater inflater,ViewGroup container,
            Bundle savedInstanceState) {
```

```
            //设置视图
            View rootView = inflater.inflate(R.layout.fragment_item_detail,
                    container,false);
            //显示内容
            if (mItem != null) {
                ((TextView) rootView.findViewById(R.id.item_detail))
                        .setText(mItem.content);
            }

            return rootView;
        }
    }
```

这个 fragment 也很简单，获取数据并显示，这里的 fragment_item_detail 的布局就是一个 TextView：

```
    <TextView
        android:id="@+id/item_detail"
    />
```

说了这么多，感觉还是有点绕。总结一下，首先 ItemListActivity 的布局为 fragment＋FrameLayout；fragment 对应的是 ItemListFragment，它是一个列表 ListFragment，布局为系统自带的简单 TextView；而 FrameLayout 在单击左边 fragment 时会被 ItemDetailFragment 替换，而这个 Fragment 的布局为 fragment_item_detail，也就是一个 TextView，如图 8-10 所示。

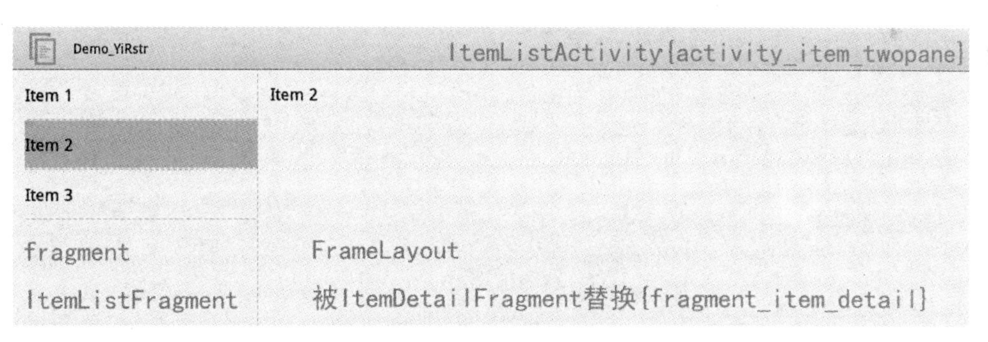

图 8-10　fragment 示意

第二节　ViewFlipper

ViewFlipper 是用来切换不同视图的一个容器，它继承自 FrameLayout。我们可以在 ViewFlipper 中添加子视图，并且通过 ViewFlipper 来控制视图之间的切换。

布局和 include 标签

Android 布局中可以使用 include 标签来使用另一个布局文件，从而达到重用的效果，如：

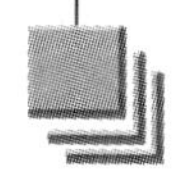

```
<RelativeLayout xmlns:android="http://schemas.android.com/apk/res/android"
    android:layout_width="fill_parent"
    android:layout_height="fill_parent" >

    <include layout="@layout/child_test"/>

</RelativeLayout>
```

child_test 的 layout 是一个简单的 TextView：

```
<?xml version="1.0" encoding="utf-8"?>
<TextView xmlns:android="http://schemas.android.com/apk/res/android"
        android:layout_width="wrap_content"
        android:layout_height="wrap_content"
        android:text="@string/hello_world"/>
```

当在一个 ViewFlipper 中有多个复杂的布局时，我们就可以使用 include 标签来将布局分层，避免代码堆积在一个布局上。

ViewFlipper 切换

ViewFlipper 可以自动进行切换，通过设置 autoStartt 以及 flipInterval（自动切换的间隔时间，单位毫秒）来实现：

```
<?xml version="1.0" encoding="utf-8"?>
<ViewFlipper xmlns:android="http://schemas.android.com/apk/res/android"
    android:layout_width="fill_parent"
    android:layout_height="fill_parent"
    android:autoStart="true"
    android:flipInterval="2000"
    android:inAnimation="@android:anim/slide_in_left"
    android:outAnimation="@android:anim/slide_out_right">

    <include layout="@layout/page_1"/>
    <include layout="@layout/page_2"/>
    <include layout="@layout/page_3"/>

</ViewFlipper>
```

此外，还可以在 Java 文件中进行切换：

```
//切换到下一个视图
public void next(View v){
    flipper.showNext();
}
//切换到前一个视图
public void prev(View v){
    flipper.showPrevious();
}
```

当然,也可以随意跳转至第 n 个视图:

```
flipper.setDisplayedChild(1);
```

第三节　MediaPlayer

MediaPlayer 生命周期

在 Android 官方网站上对 MediaPlayer 的生命周期进行了详细的说明。我在此基础上进行了标注,其中矩形框表示我们会使用到的重要方法,如图 8-11 所示。

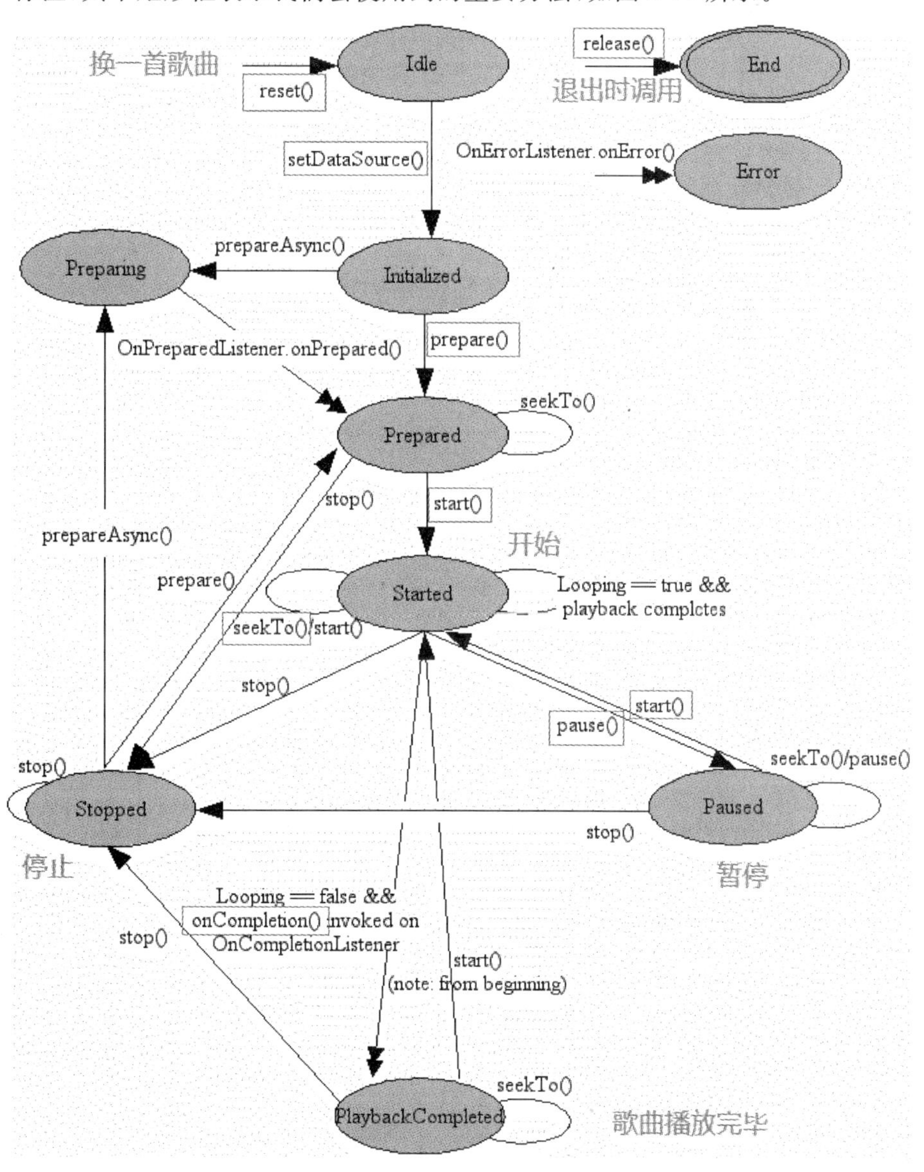

图 8-11　MediaPlayer 生命周期

播放服务实例

```
public class MusicService extends Service{

    private final IBinder binder = new MusicBinder();

    /* MediaPlayer 对象 */
    public MediaPlayer mMediaPlayer          = null;

    /* 播放列表 */
    private ArrayList<Chanson> mMusicList = new ArrayList<Chanson>();

    public ArrayList<Chanson> getMusicList() {
        return mMusicList;
    }

    /* 当前播放歌曲的索引 */
    private int currentListItme = -1;

    public class MusicBinder extends Binder {
    public MusicService getService() {
            return MusicService.this;
        }
    }
    @Override
    public IBinder onBind(Intent arg0) {
        return binder;
    }

    public void play(){
        if(currentListItme < 0){
            playMusic(0);
                    mMediaPlayer.start();
                return;
        }
        if (mMediaPlayer.isPlaying()){
            /* 暂停 */
            mMediaPlayer.pause();
        }
        else{
            /* 开始播放 */
            mMediaPlayer.start();
        }
    }

    @Override
    public void onCreate(){
        super.onCreate();
        //获取 sd 卡中的所有歌曲
    mMusicList = MusicScanner.getAllSongs(this);
    mMediaPlayer = new MediaPlayer();
    }
```

```
@Override
public void onStart(Intent intent, int id){
    super.onStart(intent, id);
}

@Override
public void onDestroy(){
    mMediaPlayer.stop();
    mMediaPlayer.release();
    super.onDestroy();
}

public void setVolume(float percent){
    mMediaPlayer.setVolume(percent, percent);
}

public void playMusic(int index)
{
currentListItme = index;
//监听器
if(onSongChangedListener != null){
    onSongChangedListener.onChange(currentListItme);
}
    try
    {
        String path = mMusicList.get(index).getUrl();
        /* 重置 MediaPlayer */
        mMediaPlayer.reset();
        /* 设置要播放的文件的路径 */
        mMediaPlayer.setDataSource(path);
        if(mMediaPlayer.isPlaying()){
            mMediaPlayer.stop();
        }
        /* 准备播放 */
        mMediaPlayer.prepare();
        /* 开始播放 */
        mMediaPlayer.start();
        //歌曲结束监听器
        mMediaPlayer.setOnCompletionListener(new OnCompletionListener() {
            public void onCompletion(MediaPlayer arg0) {
                //播放完成一首之后进行下一首
                nextMusic();
            }
        });
    }catch (Exception e){
    e.printStackTrace();
    }

}

/* 下一首 */
public void nextMusic()
{
```

```
        if (++currentListItme >= mMusicList.size()) {
            currentListItme = 0;
        }
        else {
            playMusic(currentListItme);
        }
    }

    /*上一首*/
    public void prevMusic()
    {
        if (--currentListItme >= 0) {
            currentListItme = mMusicList.size();
        }
        else {
            playMusic(currentListItme);
        }
    }

    /**
     * 将歌曲播放至指定位置
     * @param percent 百分比
     */
    public void seekTo(float percent){
        if (mMediaPlayer.isPlaying()) {
           int msec = (int) (percent * mMusicList.get(currentListItme).getDuration());
            mMediaPlayer.seekTo(msec);
        }
    }

    /**
     * @return 播放位置的百分比
     */
    public float getProgressPercent(){
        try {
            return (float)mMediaPlayer.getCurrentPosition()
                    /(float)mMediaPlayer.getDuration();
        } catch (Exception e) {
            return 0;
        }
    }
    private OnSongChangedListener onSongChangedListener;

    public void setOnSongChangedListener(OnSongChangedListener onSongChangedListener) {
        this.onSongChangedListener = onSongChangedListener;
    }
    public interface OnSongChangedListener{
    public void onChange(int index);
    }
}
```

最后一点，实现获取 sd 卡上的音频文件信息：

```
public class MusicScanner {

    public static ArrayList<Chanson> getAllSongs(Context context){
        ArrayList<Chanson> list = new ArrayList<Chanson>();
        Cursor cursor = context.getContentResolver().query(
        MediaStore.Audio.Media.EXTERNAL_CONTENT_URI,null,null,null,
        MediaStore.Audio.Media.DEFAULT_SORT_ORDER);
        while(cursor.moveToNext()){
            String tilte = cursor.getString(cursor.getColumnIndexOrThrow(
                  MediaStore.Audio.Media.TITLE));
            String artist = cursor.getString(cursor.getColumnIndexOrThrow(
                   MediaStore.Audio.Media.ARTIST));
            String url = cursor.getString(cursor.getColumnIndexOrThrow(
                   MediaStore.Audio.Media.DATA));
            int duration = cursor.getInt(cursor.getColumnIndexOrThrow(
                   MediaStore.Audio.Media.DURATION));
            list.add(new Chanson(artist,tilte,url,duration));
        }
        return list;
    }
}
```

第四节　功 能 实 现

布局实现

虽然本实例只有一个 activity，只有一个 content view，但由于布局繁多，所以 layout 文件也自然不会少，如图 8-12 所示。

```
▲ layout
    child_cart.xml
    child_menu.xml
    child_music.xml
    child_order.xml
    layout_cart.xml
    layout_details.xml
    layout_dish.xml
    layout_order.xml
    layout_song_selected.xml
    layout_song.xml
▲ layout-land
    launch.xml
```

图 8-12　布局文件

本实例固定显示横向显示，在 AndroidManifest.xml 中需要定义：

```
<application android:screenOrientation="landscape"
...
>
</application>
```

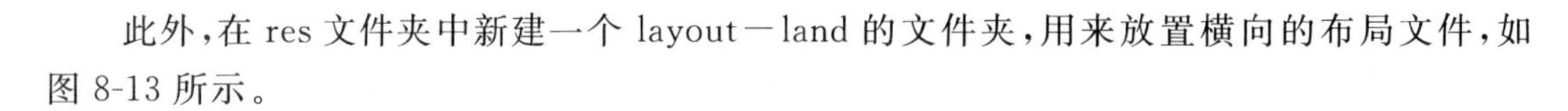

此外，在 res 文件夹中新建一个 layout－land 的文件夹，用来放置横向的布局文件，如图 8-13 所示。

layout-land
launch.xml

图 8-13 横向布局文件

布局框架的实现

launch. xml 也就是整个界面的布局框架，比较简单，如图 8-14 所示。

```
<?xml version="1.0" encoding="utf-8"?>
<RelativeLayout
    android:background="@drawable/marg">

    <!-- 菜单栏 -->
    <RadioGroup
        android:id="@+id/launch_menu_rg"
        android:orientation="vertical">

    </RadioGroup>

    <!-- 显示内容 -->
    <ViewFlipper
        android:layout_width="fill_parent"
        android:layout_height="fill_parent"
        android:layout_alignParentTop="true"
        android:layout_marginLeft="-56dip"
        android:layout_toRightOf="@+id/launch_menu_rg">

        <!-- 主页 -->
        <FrameLayout
          android:layout_width="fill_parent"
          android:layout_height="fill_parent"
              />

          <!-- 菜单 -->
          <include layout="@layout/child_menu"/>

          <!-- 购物车 -->
          <include layout="@layout/child_cart"/>

          <!-- 订单 -->
          <include layout="@layout/child_order"/>

          <!-- 音乐 -->
          <include layout="@layout/child_music"/>

    </ViewFlipper>

</RelativeLayout>
```

图 8-14　布局示意

菜单布局的实现

菜单布局 child_menu 使用了三个控件：RadioGroup、fragment 和 FrameLayout。其中 FrameLayout 主要用来占位置，在运行时被其他布局所替换，如图 8-15 所示。

```
<LinearLayout
    android:orientation = "horizontal" >

    <LinearLayout
        android:layout_weight = "1"
        android:orientation = "vertical" >
        <RadioGroup
                android:id = "@+ id/child_menu_type" >
        </RadioGroup >

        <fragment
               class = "com.mocaa.YiRstr.activity.LaunchActivity $ TitlesFragment"
            android:id = "@+ id/titles"/>
    </LinearLayout >
      <FrameLayout
          android:id = "@+ id/child_menu_details"
          android:layout_weight = "1"/>
</LinearLayout >
```

图 8-15　布局示意

其中，fragment 的适配布局为 layout_dish，是一个简单的单 RelativeLayout 结构，如图 8-16 所示。

```
<?xml version = "1.0" encoding = "utf - 8"?>
< RelativeLayout xmlns:android = "http://schemas.android.com/apk/res/android"
    android:layout_width = "fill_parent"
    android:layout_height = "fill_parent" >

    < ImageView
        android:id = "@+ id/layout_dish_img"/>

    < TextView
        android:id = "@+ id/layout_dish_price"
         />

    < TextView
        android:id = "@+ id/layout_dish_name"
         />

</RelativeLayout >
```

图 8-16　layout_dish 的布局

而右边的 FrameLayout 则会被 layout_details 替换，它的实际布局如图 8-17 所示。

```
<?xml version = "1.0" encoding = "utf - 8"?>
< RelativeLayout xmlns:android = "http://schemas.android.com/apk/res/android">

    < ImageView
        android:id = "@ + id/layout_details_img"/>

    < ImageView
        android:id = "@ + id/gap"
        android:layout_below = "@ + id/layout_details_img"
        android:layout_alignRight = "@ + id/layout_details_img"
    />

    < TableLayout
        android:id = "@ + id/layout_details_intro"
        android:layout_alignLeft = "@ + id/layout_details_img"
        android:layout_below = "@ + id/gap"
        android:shrinkColumns = "1"
        >
    </TableLayout >

    < ImageButton
        android:id = "@ + id/layout_details_cart_bt"
        android:layout_alignRight = "@ + id/layout_details_img"
        android:layout_below = "@ + id/gap"/>

</RelativeLayout >
```

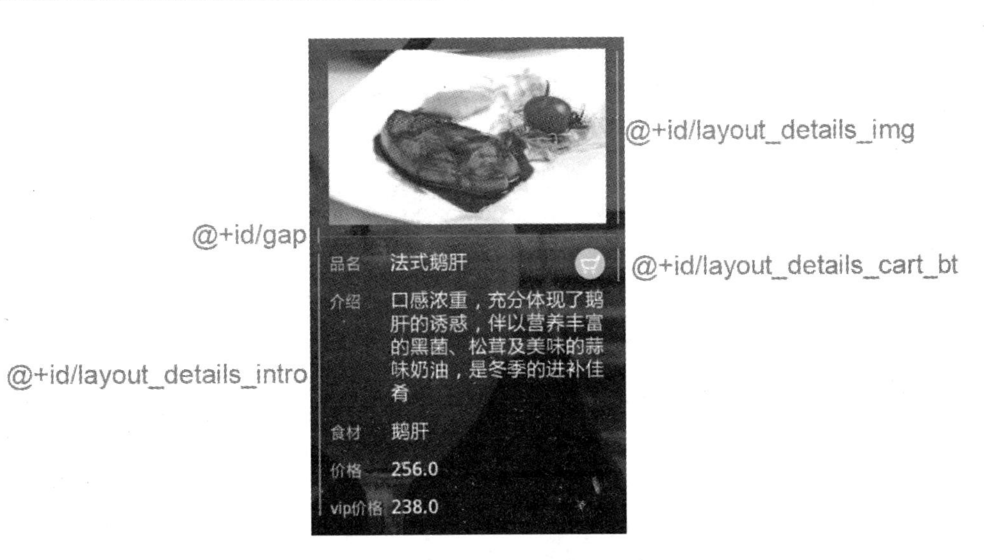

图 8-17　details_layont 布局示意

需要注意的是，菜单介绍的这个 TableLayout 中，为了避免内容超出边距，需要加上一个属性：

```
android:shrinkColumns = "1"
```

购物车布局的实现

购物车布局 child_cart 是一个简单的 RelativeLayout：

```
<?xml version = "1.0" encoding = "utf - 8"?>
<RelativeLayout xmlns:android = "http://schemas.android.com/apk/res/android"
    android:layout_width = "fill_parent"
    android:layout_height = "fill_parent"
    >
    <ListView
        android:id = "@+ id/child_cart_list"
        android:layout_above = "@+ id/cart_bottom"/>
    <LinearLayout
        android:id = "@+ id/cart_bottom" >
        <TextView
            android:id = "@+ id/child_cart_amount_tv"/>
        <Button
            android:background = "@drawable/bcg_button"
            />
    </LinearLayout>
</RelativeLayout>
```

ListView 的适配布局为 layout_cart，如图 8-18 所示。

```
<?xml version = "1.0" encoding = "utf - 8"?>
<LinearLayout xmlns:android = "http://schemas.android.com/apk/res/android"
    android:layout_width = "match_parent"
    android:layout_height = "wrap_content"
    android:paddingTop = "25dp"
    android:paddingBottom = "25dp">

    <ImageView
        android:id = "@+ id/dishIv"/>

    <RelativeLayout >
        <TextView
        android:id = "@+ id/dishTv"/>

        <LinearLayout >

            <TextView
                android:id = "@+ id/priceTv"/>

            <TextView
                android:text = " × "/>

            <ImageButton
                android:id = "@+ id/minusBt"/>
```

```
            <EditText
                android:id="@+id/countEt"/>

            <ImageButton
                android:id="@+id/plusBt"/>
        </LinearLayout>

        <ImageButton
                android:id="@+id/deleteBt"/>

    </RelativeLayout>
</LinearLayout>
```

图 8-18　layout_cart 布局示意

订单布局的实现

订单的布局与购物车的布局类似，只是少了一个按钮：

```
<?xml version="1.0" encoding="utf-8"?>
<RelativeLayout xmlns:android="http://schemas.android.com/apk/res/android"
    android:layout_width="fill_parent"
    android:layout_height="fill_parent"
    >
    <ListView
        android:id="@+id/child_order_list"
        android:layout_above="@+id/child_order_amount_tv"/>
    <TextView
        android:id="@+id/child_order_amount_tv"
        android:layout_alignParentBottom="true"/>
</RelativeLayout>
```

同样地，它的布局与购物车类似，在这里就不细讲了，如图 8-19 所示。

```
<?xml version="1.0" encoding="utf-8"?>
<LinearLayout xmlns:android="http://schemas.android.com/apk/res/android"
```

```
    android:layout_width = "match_parent"
    android:layout_height = "wrap_content" >

    < ImageView
        android:id = "@+ id/dishIv"/>

    < RelativeLayout >
        < TextView
            android:id = "@+ id/dishTv"/>

        < LinearLayout
            android:layout_alignParentBottom = "true"
            android:layout_alignLeft = "@+ id/dishTv" >

            < TextView
                android:id = "@+ id/priceTv"/>
            < TextView
                android:text = " × "/>
            < TextView
                android:id = "@+ id/countTv"/>

        </LinearLayout >

        < TextView
            android:id = "@+ id/amountTv"
            android:layout_alignParentRight = "true"/>

    </RelativeLayout >
</LinearLayout >
```

图 8-19　订单布局效果

音乐播放器布局的实现

音乐播放器的布局较为繁琐，所以我们先来看图 8-20 音乐播放器布局示意图：

第一层为 RelativeLayout，也就是根。

第二层有三个元素：进度条(@+id/music_progress_bar)，控制栏(@+id/group_control_menu)和播放列表(@+id/music_list)。

第三层为控制栏层，包括三个元素，它们都是 LinearLayout：歌曲信息(@+id/group_music_info)，按钮(@+id/group_play_button)以及音量控制栏(@+id/group_volume_control)。

第四层以及第五层为歌曲信息的子层，我们不用管它。

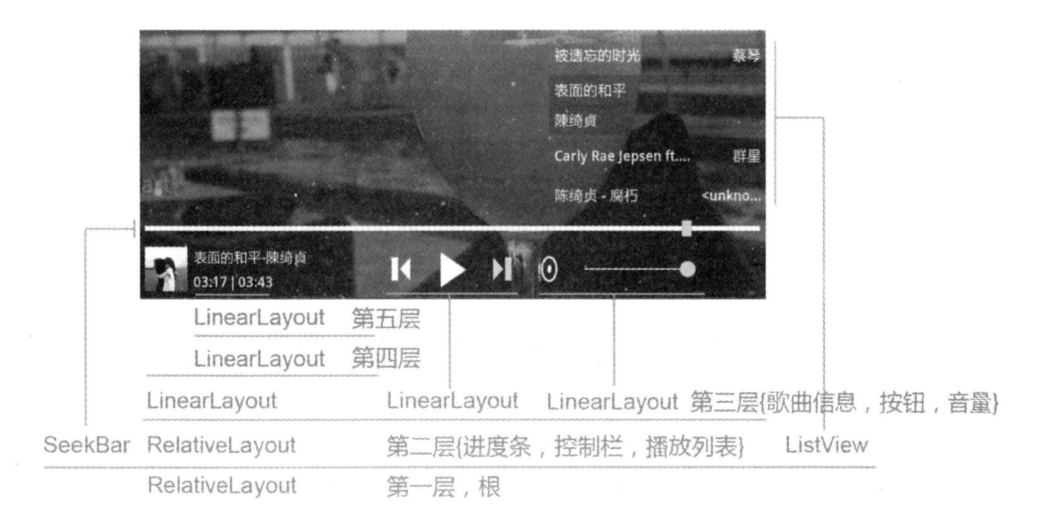

图 8-20 音乐播放器布局

代码框架如下：

```
<?xml version = "1.0" encoding = "utf - 8"?>
< RelativeLayout xmlns:android = "http://schemas.android.com/apk/res/android"
    android:id = "@+ id/child_music_root">
    < ListView
            android:id = "@+ id/music_list"
            android:layout_alignParentRight = "true"
            android:layout_above = "@+ id/music_progress_bar"/>
    < SeekBar
            android:id = "@+ id/music_progress_bar"
            android:layout_above = "@+ id/group_control_menu"/>
    < RelativeLayout
            android:id = "@+ id/group_control_menu"
            android:layout_alignParentBottom = "true" >
            < LinearLayout
                android:id = "@+ id/group_music_info"
                android:layout_toLeftOf = "@+ id/group_play_button" >
                < ImageView
                    android:id = "@+ id/music_artist_img"/>
                < LinearLayout
                    android:orientation = "vertical"
                    >
                    < TextView
                        android:id = "@+ id/music_title_tv"/>
                    < LinearLayout
                        android:gravity = "bottom"
                        android:layout_weight = "1" >
                        < TextView
                            android:id = "@+ id/music_time_tv"/>
                        < TextView
                            android:id = "@+ id/music_total_tv"/>
```

```
                </LinearLayout>
            </LinearLayout>
        </LinearLayout>
        <LinearLayout
            android:id="@+id/group_play_button">

            <Button
                android:id="@+id/music_prev_bt"/>
            <CheckBox
                android:id="@+id/music_play_bt"/>
            <Button
                android:id="@+id/music_next_bt"/>

    </LinearLayout>

    <LinearLayout
        android:id="@+id/group_volume_control"
        android:layout_toRightOf="@+id/group_play_button">
        <CheckBox
            android:id="@+id/music_sound_bt"/>
        <SeekBar
            android:id="@+id/music_sound_bar"/>
    </LinearLayout>
</RelativeLayout>

</RelativeLayout>
```

在播放列表中，播放中的曲目布局与未播放曲目布局是不一样的，如图 8-21 所示，在适配器中通过判断来实现。

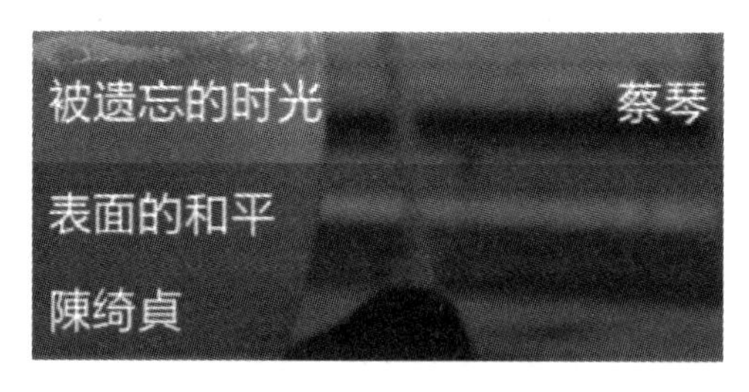

图 8-21　播放中曲目

菜单、购物车、订单功能实现

可以说这部分的功能很常规，通过 Activity 和 Adapter 之间进行交互来展示数据。在这之前，我们先定义一个全局变量，用于保存菜单、购物车以及订单数据：

```
public class Dishes extends Application{

    //所有的菜单
    private ArrayList<Dish> allDishes;
```

```
//购物车
private HashSet < OrderDish > shoppingCart = new HashSet < OrderDish >();
//订单
private HashSet < OrderDish > order = new HashSet < OrderDish >();

public ArrayList < Dish > getAllDishes() {
    return allDishes;
}
public void setAllDishes(ArrayList < Dish > allDishes) {
    this.allDishes = allDishes;
}
public HashSet < OrderDish > getShoppingCart() {
    return shoppingCart;
}
public HashSet < OrderDish > getOrders() {
    return order;
}
private float getAmount(HashSet < OrderDish > dishes){
    float amount = 0f;
    Object[] objOrder = dishes.toArray();
    for(int i = 0; i < objOrder.length; i++){
        OrderDish od = (OrderDish)objOrder[i];
        String no = od.getNoDish();
        Dish dish = findDishByNo(no);
        amount + = od.getDishNum() * dish.getPriceMarket();
    }
    return amount;
}
//获取购物车总价
public float getShoppingCartAmount(){
    return getAmount(shoppingCart);
}
//获取订单总价
public float getOrdersAmount(){
    return getAmount(order);
}
public Dish findDishByNo(String no){
        for(int i = 0; i < allDishes.size(); i++){
            if(allDishes.get(i).getNo().equals(no)){
                return allDishes.get(i);
            }
        }
        return null;
}
public ArrayList < Dish > getDishesByType(int type){
        ArrayList < Dish > dishes = new ArrayList < Dish >();
        for(int i = 0; i < allDishes.size(); i++){
            if(allDishes.get(i).getType() == type){
                dishes.add(allDishes.get(i));
            }
```

```
            }
            return dishes;
    }
    //下单
    public void pay(){
            order.clear();
            order.addAll(shoppingCart);
            shoppingCart.clear();
    }
}
```

在 Activity 中，进行了数据加载以及各个模块的初始化：

```
@Override
    public void onCreate(Bundle savedInstanceState) {
        super.onCreate(savedInstanceState);
        setContentView(R.layout.launch);

        //全局变量
    dishes = (Dishes) getApplication();
    /*
     * 加载所有菜单数据
     * 在实际项目中,这个模块应该是从网络上获取,
     * 或者是从本地读取,定时与服务端同步
     */
    loadAllDishes();
    //初始化界面框架 ViewFlipper 与 RadioGroup
    initFramework();
    //初始化购物车
    initShoppingCart();
    //初始化订单
    initOrder();
    //初始化音乐播放器
    initMusicPlayer();
}
```

在初始化界面框架中的功能为设定事件，选择第 n 个切换到第 n 个界面：

```
mFlipper = (ViewFlipper) findViewById(R.id.viewFlipper);
mRadioGroup = (RadioGroup) findViewById(R.id.launch_menu_rg);
mRadioGroup.setOnCheckedChangeListener(new OnCheckedChangeListener(){
    @Override
    public void onCheckedChanged(RadioGroup group, int id) {
        switch(id){
        case R.id.menu_home:
            mFlipper.setDisplayedChild(0);
            break;
```

购物车的初始化：

```
ListView cart = (ListView) findViewById(R.id.child_cart_list);
TextView amount_cart = (TextView) findViewById(R.id.child_cart_amount_tv);
cartAdapter = new CartListAdapter(this,amount_cart);
cart.setAdapter(cartAdapter);
//选择大类(西餐、中餐、甜品、汤、酒水)
RadioGroup types = (RadioGroup) findViewById(R.id.child_menu_type);
    types.setOnCheckedChangeListener(new OnCheckedChangeListener() {
        @Override
        public void onCheckedChanged(RadioGroup arg0, int id) {
            switch(id){
            case R.id.type_chinese:
                //获得类别,更新列表
                getDishes(Dish.TYPE_CHINESE);
                break;
            case R.id.type_western:
```

订单模块初始化：

```
ListView order = (ListView) findViewById(R.id.child_order_list);
TextView amount_order = (TextView) findViewById(R.id.child_order_amount_tv);
OrderListAdapter adapter = new OrderListAdapter(this,amount_order);
order.setAdapter(adapter);
```

音乐模块的具体实现被封装在了 MusicChild 中,这边只调用了它的方法。具体实现会在下一节中讲解。

```
child_music = new MusicChild();
child_music.create(findViewById(R.id.child_music_root),this);
```

最后一部分,菜单的实现,它是一个教科书式的 fragment：

```
public static class TitlesFragment extends ListFragment {

    boolean mDualPane;
    int mCurCheckPosition = 0;
    static FragmentLayout adpter;

    @Override
    public void onActivityCreated(Bundle savedInstanceState) {
        super.onActivityCreated(savedInstanceState);

        ListView listView = getListView();
        //避免在滚动时出现白色边
        listView.setCacheColorHint(Color.TRANSPARENT);
        adapter = new FragmentLayout(dishList,this.getActivity());
        setListAdapter(adapter);

        View detailsFrame = getActivity().findViewById(R.id.child_menu_details);
```

```
        mDualPane = detailsFrame != null
            && detailsFrame.getVisibility() == View.VISIBLE;

        if (savedInstanceState != null) {
            /从保存的状态中取出数据
            mCurCheckPosition = savedInstanceState.getInt("curChoice",0);
        }

        if (mDualPane) {
            listView.setChoiceMode(ListView.CHOICE_MODE_SINGLE);
            showDetails(mCurCheckPosition);
        }
    }

    @Override
    public void onSaveInstanceState(Bundle outState) {
        super.onSaveInstanceState(outState);
        outState.putInt("curChoice",mCurCheckPosition);              //保存当前的下标
    }

    @Override
    public void onListItemClick(ListView l,View v,int position,long id) {
        super.onListItemClick(l,v,position,id);
        showDetails(position);
    }

    void showDetails(int index) {
        mCurCheckPosition = index;
        if (mDualPane) {
            getListView().setItemChecked(index,true);
            DetailsFragment details = (DetailsFragment) getFragmentManager()
                .findFragmentById(R.id.child_menu_details);
            if (details == null || details.getShownIndex() != index) {
                details = DetailsFragment.newInstance(mCurCheckPosition);

                //得到一个 fragment 事务(类似 sqlite 的操作)
                FragmentTransaction ft = getFragmentManager()
                    .beginTransaction();
                //将得到的 fragment 替换当前的 viewGroup 内容,add 则不替换会依次累加
                ft.replace(R.id.child_menu_details,details);
                ft.setTransition(FragmentTransaction.TRANSIT_FRAGMENT_FADE);
                                                                //设置动画效果
                ft.commit();                                    //提交
            }
        }
    }
}
/**
 *作为界面的一部分,为 fragment 提供一个 layout
 */
```

```
public static class DetailsFragment extends Fragment {

    public static DetailsFragment newInstance(int index) {
        DetailsFragment details = new DetailsFragment();
        Bundle args = new Bundle();
        args.putInt("index", index);
        details.setArguments(args);
        return details;
    }

    public int getShownIndex() {
        return getArguments().getInt("index", -1);
    }

    @Override
    public View onCreateView(LayoutInflater inflater, ViewGroup container,
        Bundle savedInstanceState) {
        if (container == null)
              return null;
        View view = LayoutInflater.from(getActivity())
              .inflate(R.layout.layout_details, null);
        ImageButton cartBt = (ImageButton) view.findViewById(
              R.id.layout_details_cart_bt);
        cartBt.setOnClickListener(new OnClickListener(){
            @Override
            public void onClick(View arg0) {
                int index = getShownIndex();
                Dish dish = dishList.get(index);
                OrderDish order = new OrderDish(dish.getNo());
                cartAdapter.addToCart(order);
            }

        });
        ImageView img = (ImageView) view.findViewById(R.id.layout_details_img);
        TextView name = (TextView) view.findViewById(
            R.id.layout_details_name_tv);
        TextView intro = (TextView) view.findViewById(
            R.id.layout_details_intro_tv);
        TextView expl = (TextView) view.findViewById(
            R.id.layout_details_expl_tv);
        TextView price = (TextView) view.findViewById(
            R.id.layout_details_price_tv);
        TextView vip_price = (TextView) view.findViewById(
            R.id.layout_details_vip_price_tv);

        if(getShownIndex() >= 0){
            Dish dish = dishList.get(getShownIndex());
            img.setImageResource(dish.getImgId());
            name.setText(dish.getName());
            intro.setText(dish.getInto());
```

```
                expl.setText(dish.getExplanation());
                price.setText(dish.getPriceMarket() + "");
                vip_price.setText(dish.getPriceVIP() + "");
            }
            return view;
        }
}
```

音乐播放器的实现

在第三大节 MusicService 的基础上，本节来介绍音乐播放器的实现。在这里出现的 MusicService 就是第三节播放器实例中讲解的 MusicService，因此在此不多赘述。MusicChild 负责实现界面上的更新，而 MusicService 负责播放功能的实现。在 MusicChild 中定义几个函数：

```
public void create(View v,Context context){
    this.context = context;
    this.root = v;
    //绑定并开始服务
    serviceIntent = new Intent(context,MusicService.class);
    context.startService(serviceIntent);
    context.bindService(serviceIntent,mConnection,0);
}
//开始进度条以及播放时间的线程
public void start(){
    new ProgressThread().start();
    new TimeThread().start();
}
//结束线程
public void stop(){
        mIsCreated = false;
}
public void destroy(){
    mIsCreated = false;
    //停止服务
    context.unbindService(mConnection);
    context.stopService(serviceIntent);
}
```

在绑定回调函数中定义初始化代码：

```
private ServiceConnection mConnection = new ServiceConnection() {
    //回调方法，当调用 bindService 时回调
    public void onServiceConnected(ComponentName className, IBinder localBinder)
    {
        //获取 service 对象
        mService = ((MusicBinder)localBinder).getService();
```

```
            init();
        }
        //回调方法,当调用 unbindService 时回调
        public void onServiceDisconnected(ComponentName arg0) {
            mService = null;
        }
    };
    public void init(){

        mService.setOnSongChangedListener(new OnSongChangedListener() {
            @Override
            public void onChange(int index) {
                //开始播放第 index 首歌曲
                mSelectedIndex = index;
                Chanson song = mService.getMusicList().get(index);
                //设置歌曲信息
                titleTv.setText(song.getTitle() + " - " + song.getArtist());
                totalTv.setText(" | " + DurationUtil.getTime(song.getDuration()));
                //在播放列表中选中该歌曲
                mSongAdapter.select(index);
                //设置为暂停按钮
                mPlayBt.setChecked(false);
            }
        });
        mPrevBt = (Button) root.findViewById(R.id.music_prev_bt);
        mPlayBt = (CheckBox) root.findViewById(R.id.music_play_bt);
        mNextBt = (Button) root.findViewById(R.id.music_next_bt);
        mListView = (ListView) root.findViewById(R.id.music_list);
        mVolumeBox = (CheckBox) root.findViewById(R.id.music_sound_bt);
        //静音/还原
        mVolumeBox.setOnCheckedChangeListener(new OnCheckedChangeListener() {
            @Override
            public void onCheckedChanged(CompoundButton arg0,boolean flag) {
                if(flag){
                    mService.setVolume(0);
                }else{
                    mService.setVolume((float)pref.getVolume()/MAX_VOLUME);
                }
            }
        });
        mProgressBar = (SeekBar) root.findViewById(R.id.music_progress_bar);
        MAX_PROGRESS = mProgressBar.getMax();
        mProgressBar.setOnSeekBarChangeListener(new OnSeekBarChangeListener() {
            @Override
            public void onStopTrackingTouch(SeekBar seekBar) {
                //歌曲播放至
                float progress = seekBar.getProgress();
                mProgressPercent = progress/MAX_PROGRESS;
                mService.seekTo(mProgressPercent);
                //继续进度条的线程
```

```
            mIsSeeking = true;
        }

        @Override
        public void onStartTrackingTouch(SeekBar seekBar) {
            //暂停设置进度条的线程
            mIsSeeking = false;
        }

        @Override
        public void onProgressChanged(SeekBar seekBar, int progress,
                boolean fromUser) {
        }
    });

    //保存用户信息,如音量
    pref = new UserPref(context);
    mVolumeBar = (SeekBar) root.findViewById(R.id.music_sound_bar);
    MAX_VOLUME = mVolumeBar.getMax();
    mVolumeBar.setProgress(pref.getVolume());
    mVolumeBar.setOnSeekBarChangeListener(new OnSeekBarChangeListener() {

        @Override
        public void onStopTrackingTouch(SeekBar seekBar) {
        }

        @Override
        public void onStartTrackingTouch(SeekBar seekBar) {
        }

        @Override
        public void onProgressChanged(SeekBar seekBar, int progress,
                boolean fromUser) {
            //设置音量
            pref.setVolume(progress);
            mService.setVolume((int)progress/MAX_VOLUME);
            mVolumeBox.setChecked(false);
        }
    });

    mSongAdapter = new SongListAdapter(context, mService.getMusicList());
    mListView.setAdapter(mSongAdapter);
    mListView.setOnItemClickListener(new OnItemClickListener() {
        @Override
        public void onItemClick(AdapterView<?> arg0, View arg1, int index,
                long arg3) {
            //播放歌曲
            mSelectedIndex = index;
            mService.playMusic(index);
            mPlayBt.setChecked(false);
        }
    });
```

```
    mPlayBt.setOnClickListener(new OnClickListener(){
        public void onClick(View view){
            mService.play();
        }
    });

    //下一首
    mNextBt.setOnClickListener(new OnClickListener(){
            @Override
            public void onClick(View arg0){
                mService.nextMusic();
            }
    });
    //上一首
    mPrevBt.setOnClickListener(new OnClickListener(){
            @Override
            public void onClick(View arg0){
                mService.prevMusic();
            }
    });

    artistImg = (ImageView) root.findViewById(R.id.music_artist_img);
    titleTv = (TextView) root.findViewById(R.id.music_title_tv);
    timeTv = (TextView) root.findViewById(R.id.music_time_tv);
    totalTv = (TextView) root.findViewById(R.id.music_total_tv);
}
```

控制进度条以及播放时间的线程：

```
//进度条的线程
class ProgressThread extends Thread{
    public void run(){
        while(mIsCreated){
            while(mIsSeeking){
                mProgressPercent = mService.getProgressPercent();
                mProgressBar.setProgress(
                    (int) (mProgressPercent * MAX_PROGRESS));
                try {
                    sleep(100);
                } catch (InterruptedException e) {
                    e.printStackTrace();
                }
            }
        }
    }
};
//播放时间的线程
class TimeThread extends Thread{
    public void run(){
```

```
            while(mIsCreated){
                if(mSelectedIndex >= 0){
                    int msec = (int) (mProgressPercent *
                        mService.getMusicList().get(mSelectedIndex).getDuration());
                    Message msg = new Message();
                    //将毫秒转换为分钟显示 XX:XX
                    msg.obj = DurationUtil.getTime(msec);
                    handler.sendMessage(msg);
                }
                try {
                    sleep(1000);
                } catch (InterruptedException e) {
                    e.printStackTrace();
                }
            }
        }
    };
    private Handler handler = new Handler(){
        @Override
        public void handleMessage(Message msg){
            String time = (String) msg.obj;
            timeTv.setText(time);
        }
    };
```

第九章　Android 4. x初探

What's new?

清单

Demo 代码：

\demo\Demo_Douban

Target SDK：

Android 4.2

第一节　Android 4. x 的标准化框架

新建一个工程

Android 4.0 提供了一套标准化的界面框架，这个框架使用在了 Google 的多款 Android 产品中，如图 9-1 所示。

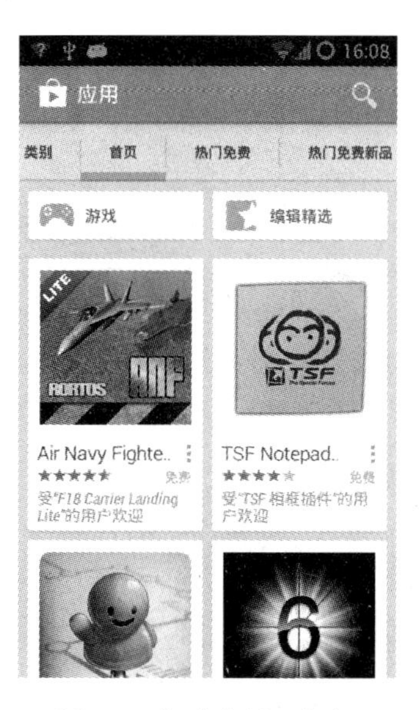

图 9-1　标准化界面框架

首先，新建一个工程，以4.2作为Target SDK，如图9-2所示。

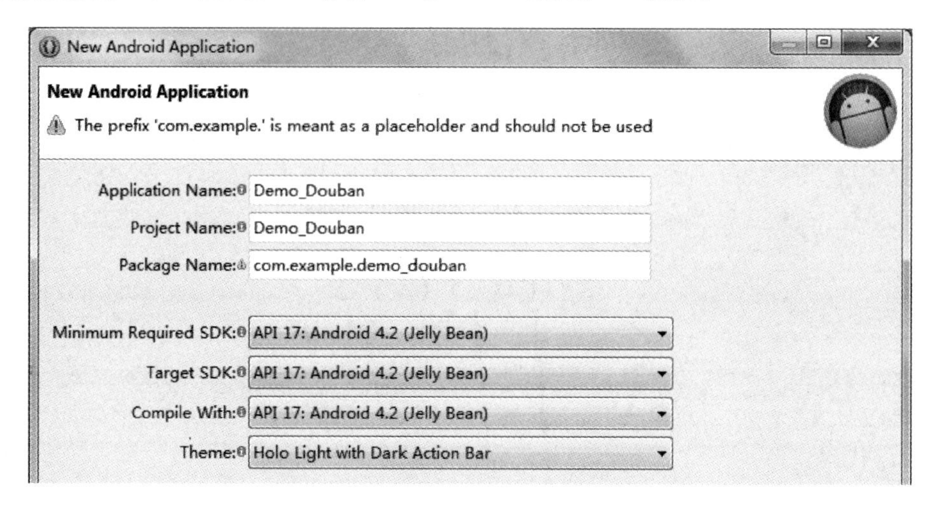

图 9-2　选择 Target SDK

选择 Blank Activity，如图9-3所示。

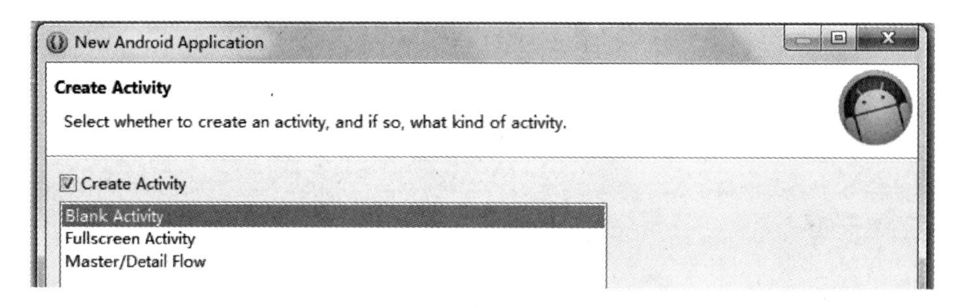

图 9-3　选择 Blank Activity

选择 Scrollable Tabs+Swipe，如图9-4所示。

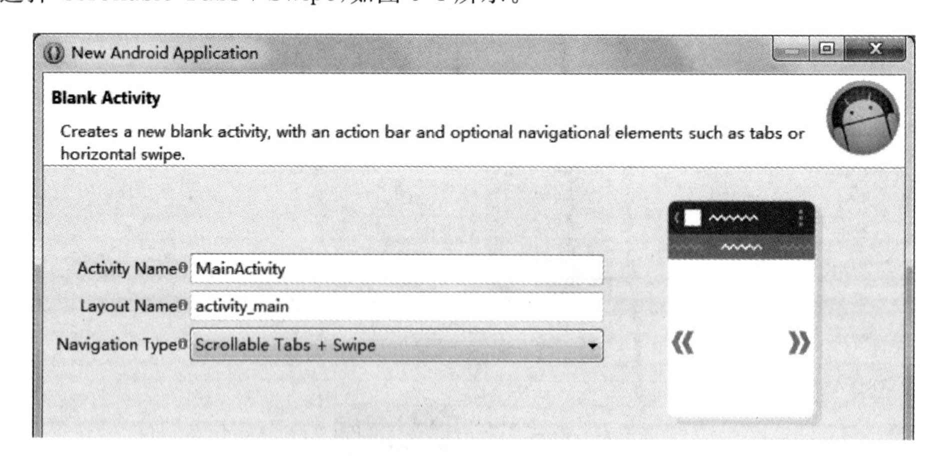

图 9-4　选择 Scrollable Tabs+Swipe

生成的代码包含一个Activity和两个layout。在有了fragment的基础之后，这部分的内容也就比较简单了。我们先来看activity_main.xml的代码：

```
<android.support.v4.view.ViewPager xmlns:android="http://schemas.android.com/apk/res/android"
    xmlns:tools="http://schemas.android.com/tools"
    android:id="@+id/pager"
    android:layout_width="match_parent"
    android:layout_height="match_parent"
    tools:context=".MainActivity">

    <android.support.v4.view.PagerTitleStrip
        android:id="@+id/pager_title_strip"
        android:layout_width="match_parent"
        android:layout_height="wrap_content"
        android:layout_gravity="top"
        android:background="#33b5e5"
        android:paddingBottom="4dp"
        android:paddingTop="4dp"
        android:textColor="#fff"/>

</android.support.v4.view.ViewPager>
```

ViewPager 和 PagerTitleStrip

ViewPager 是 ViewFlipper 的进化版，也是用于叠放不同的视图，但是可以支持视图的拖动。这个控件被使用在了大多数的应用上，如图 9-5 所示。

图 9-5　ViewPager 视图

PaperTitleStrip 也就是 ViewPager 的标题，只能作为 ViewPager 的子控件出现，如图 9-6 所示，在 Java 代码中通过重写 getPageTitle(int)方法来设置标题。

图 9-6　ViewPager 的 title

在生成的 MainActivity 中设置适配器：

```
@Override
protected void onCreate(Bundle savedInstanceState) {
    super.onCreate(savedInstanceState);
    setContentView(R.layout.activity_main);

    //设置适配器
    mSectionsPagerAdapter = new SectionsPagerAdapter(
            getSupportFragmentManager());
    mViewPager = (ViewPager) findViewById(R.id.pager);
    mViewPager.setAdapter(mSectionsPagerAdapter);

}
```

适配器的作用在于：①设置标题；②显示子视图。

```
public class SectionsPagerAdapter extends FragmentPagerAdapter {

    public SectionsPagerAdapter(FragmentManager fm) {
        super(fm);
    }

    @Override
    public Fragment getItem(int position) {
        //获取在 position 位置上的 fragment
        Fragment fragment = new DummySectionFragment();
        Bundle args = new Bundle();
        args.putInt(DummySectionFragment.ARG_SECTION_NUMBER,position + 1);
        fragment.setArguments(args);
        return fragment;
    }

    @Override
    public int getCount() {
        //显示 3 个子窗口
        return 3;
    }

    //获取标题
    @Override
    public CharSequence getPageTitle(int position) {
        Locale l = Locale.getDefault();
        switch (position) {
        case 0:
            return getString(R.string.title_section1).toUpperCase(l);
        case 1:
            return getString(R.string.title_section2).toUpperCase(l);
```

```
            case 2:
                return getString(R.string.title_section3).toUpperCase(l);
            }
            return null;
        }
    }
```

子视图的显示使用了 fragment，这个 fragment 为：

```
public static class DummySectionFragment extends Fragment {

    public static final String ARG_SECTION_NUMBER = "section_number";

    public DummySectionFragment() {
    }

    @Override
    public View onCreateView(LayoutInflater inflater,ViewGroup container,
            Bundle savedInstanceState) {
        View rootView = inflater.inflate(R.layout.fragment_main_dummy,
                container,false);
        TextView dummyTextView = (TextView) rootView
                .findViewById(R.id.section_label);
        dummyTextView.setText(Integer.toString(getArguments().getInt(
                ARG_SECTION_NUMBER)));
        return rootView;
    }
}
```

使用 Action Bar 和 Navigation

我们在布局中添加一个跳转到另一个 Activity 的按钮，第二个 Activity 代码如下：

```
public class DetailActivity extends Activity {

    @Override
    protected void onCreate(Bundle savedInstanceState) {
        super.onCreate(savedInstanceState);
        setContentView(R.layout.activity_detail);

        //显示返回的图标
        getActionBar().setDisplayHomeAsUpEnabled(true);
    }

    @Override
    public boolean onOptionsItemSelected(MenuItem item) {
            //返回到 MainActivity
        switch (item.getItemId()) {
        case android.R.id.home:
```

```
                NavUtils.navigateUpTo(this,
                        new Intent(this,MainActivity.class));
                return true;
            }
            return super.onOptionsItemSelected(item);
        }
    }
```

在这个新的 Activity 中，会出现这样一个返回的指示，单击它之后，返回到我们指定的 Activity，也就是 MainActivity，如图 9-7 所示。

图 9-7　返回的指示按钮

第二节　第六人 GridLayout

我们在前面介绍 Android 的五大布局时提到过，自 Android 4.0 之后增加了第六个布局 GridLayout。那么就来看看这个布局是怎样使用的。

GridLayout 江湖人称网格布局，就是把整个界面分为 n 行 m 列，然后再用控件来填充。在没有 GridLayout 时，我们需要使用 TableLayout 来实现，但是明显比较麻烦，要定义多个 TableRow。

GridLayout 的一个经典例子就是计算器如图 9-8 所示。我们用 GridLayout 嵌套 Button 来实现一个简单的计算器布局。

图 9-8　计算器

```
<?xml version = "1.0" encoding = "utf - 8"?>
<LinearLayout xmlns:android = "http://schemas.android.com/apk/res/android"
```

```
    android:layout_width = "fill_parent"
    android:layout_height = "fill_parent"
    android:gravity = "center"
    >
    < GridLayout
        android:layout_width = "wrap_content"
        android:layout_height = "wrap_content"
        android:rowCount = "4"
        android:columnCount = "5" >
        < Button
            android:text = "7"/>
        < Button
            android:text = "8"/>
        < Button
            android:text = "9"/>
        < Button
            android:text = "/"/>
        < Button
            android:text = " % "/>
        < Button
            android:text = "4"/>
        < Button
            android:text = "5"/>
        < Button
            android:text = "6"/>
        < Button
            android:text = " × "/>
        < Button
            android:text = "1/x"/>
        < Button
            android:text = "1"/>
        < Button
            android:text = "2"/>
        < Button
            android:text = "3"/>
        < Button
            android:text = " - "/>
        < Button
            android:layout_rowSpan = "2"
            android:layout_gravity = "fill_vertical"
            android:text = " = "/>
        < Button
            android:layout_columnSpan = "2"
            android:layout_gravity = "fill_horizontal"
            android:text = "0"/>
        < Button
            android:text = "."/>
        < Button
            android:text = " + "/>
    </GridLayout >
</LinearLayout >
```

效果如图 9-9 所示。

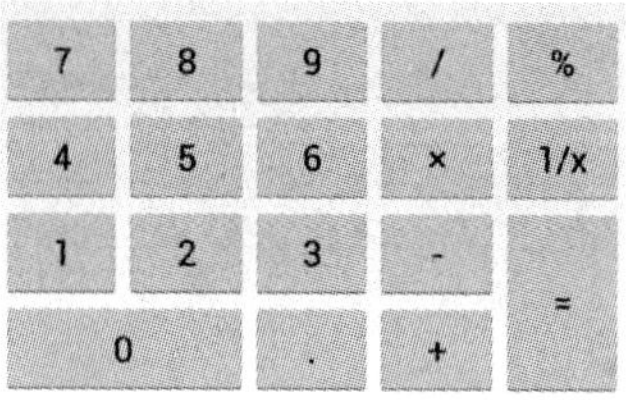

图 9-9　自制计算机器视图

在 GridLayout 中定义行数以及列数：

```
android:rowCount = "4"
android:columnCount = "5"
```

这表示这个 GridLayout 的空间为 4 行 5 列，它的子控件就会依照从左上到右下的方式依次填充整个布局。

如果要设置一个控件占多少个格子，可以使用：

```
android:layout_rowSpan = "2"
android:layout_columnSpan = "2"
```

稍微修改一下代码，把“=”的 layout_gravity 改为：

```
android:layout_gravity = "center"
```

结果如图 9-10 所示。

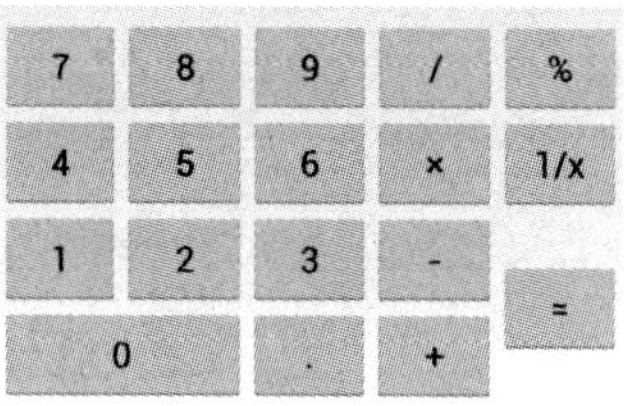

图 9-10　修改后的结果

我们可以看到，“=”这个按钮只有一个格子的大小，居中于第 3 第 4 行中间。这是因为 layout_columnSpan 这个属性只是给控件设定一个 parent，而这个控件可以通过 layout_gravity 来设定自己相对于父亲的位置。

第三节　增强 Notification

我们早在第 4 章就讲过 Notification。在 Android 4.0 之后，Notification 又增加了新的功能，主要表现为增加了 Notification 的样式，支持自定义的 Notification，如图 9-11 所示。

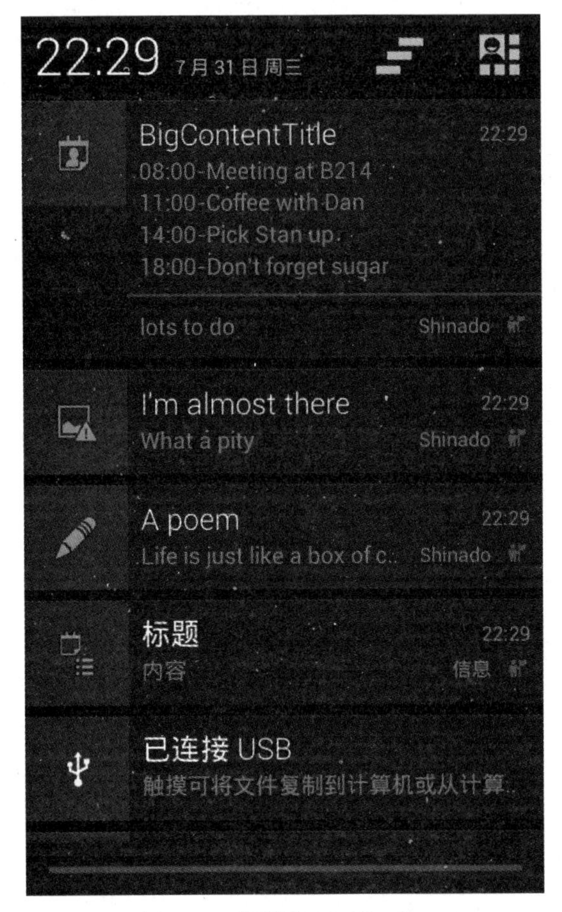

图 9-11　增强的 Notification

首先，一个正常的 Notification 包含了图标、标题、内容、信息以及小图标，如图 9-12 所示。

图 9-12　标准 Notification

这个 Notification 的代码为：

```
public void noti_normal(View v) {
    Bitmap icon = BitmapFactory.decodeResource(getResources(),
            android.R.drawable.ic_menu_agenda);
    Notification notification = new NotificationCompat.Builder(this)
                .setLargeIcon(icon)
                .setSmallIcon(smallIcon)
                .setContentInfo("信息")
                .setContentTitle("标题")
                .setContentText("内容")
                .setAutoCancel(true)
                .setDefaults(Notification.DEFAULT_ALL)
```

```
            .build();
        manager.notify(1,notification);
    }
```

Android 自备了三种特殊的 Notification，首先是支持长文字的 Notification，如图 9-13 所示。

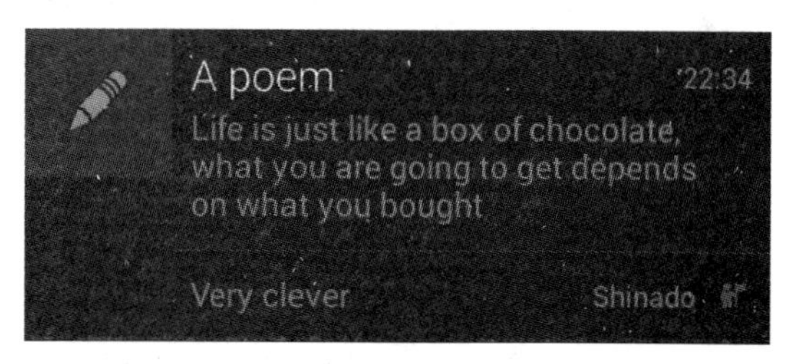

图 9-13　长文字 Notification

代码的关键是创建了一个 BigTextStyle：

```
public void noti_big_text(View v) {
    Bitmap icon = BitmapFactory.decodeResource(getResources(),
            android.R.drawable.ic_menu_edit);
    String msg = "Life is just like a box of chocolate," +
            "what you are going to get depends on what you bought";
    NotificationCompat.BigTextStyle textStyle = new BigTextStyle();
    textStyle.setBigContentTitle("A poem")
            .setSummaryText("Very clever")
            .bigText(msg);
    Notification notification = new NotificationCompat.Builder(this)
            .setLargeIcon(icon)
            .setSmallIcon(smallIcon)
            .setContentInfo("Shinado")
            .setContentTitle("A poem")
            .setContentText(msg)
            .setStyle(textStyle)
            .setAutoCancel(true)
            .setDefaults(Notification.DEFAULT_ALL)
            .build();
    manager.notify(2,notification);
}
```

同时，当这个 Notification 处于非第一条的状态时，它显示的内容就跟普通 Notification 一致，如图 9-14 所示。

图 9-14　省略状态下的 Notification

大图片的 Notification，如图 9-15 所示。

```
public void noti_big_pic(View v) {
    Bitmap icon = BitmapFactory.decodeResource(getResources(),
            android.R.drawable.ic_menu_report_image);
    Bitmap pic = BitmapFactory.decodeResource(getResources(),
            R.drawable.pic);
    NotificationCompat.BigPictureStyle pictureStyle = new BigPictureStyle();
    pictureStyle.setBigContentTitle("I'm almost there")
            .setSummaryText("What a pity")
            .bigPicture(pic);
    Notification notification = new NotificationCompat.Builder(this)
            .setLargeIcon(icon)
            .setSmallIcon(smallIcon)
            .setContentInfo("Shinado")
            .setContentTitle("I'm almost there")
            .setContentText("What a pity")
            .setStyle(pictureStyle)
            .setAutoCancel(true).setDefaults(Notification.DEFAULT_ALL)
            .build();
    manager.notify(3,notification);
}
```

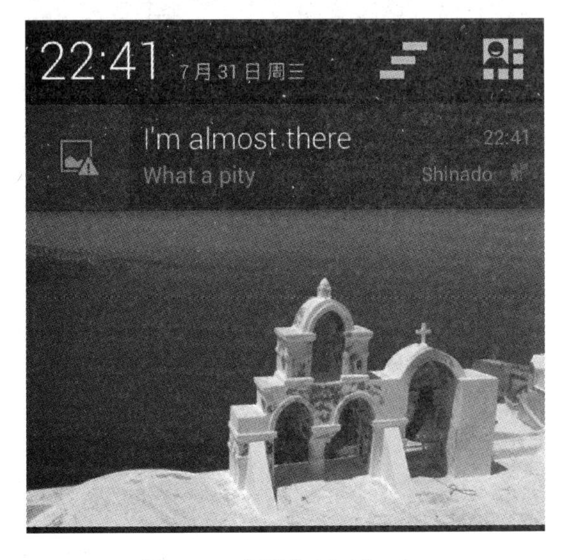

图 9-15　大图片 Notification

最后一种，收件箱模式，其实也是一种 BitTextStyle，如图 9-16 所示。

```
public void noti_inbox(View v) {
    Bitmap icon = BitmapFactory.decodeResource(getResources(),
            android.R.drawable.ic_menu_my_calendar);
    NotificationCompat.InboxStyle inboxStyle =
        new NotificationCompat.InboxStyle();
```

```
        inboxStyle.setBigContentTitle("4 todos today")
            .setSummaryText("lots to do");
        inboxStyle.addLine("08:00 - Meeting at B214");
        inboxStyle.addLine("11:00 - Coffee with Dan");
        inboxStyle.addLine("14:00 - Pick Stan up");
        inboxStyle.addLine("18:00 - Don't forget sugar");
        Notification notification = new NotificationCompat.Builder(this)
            .setLargeIcon(icon).setSmallIcon(smallIcon)
            .setContentInfo("Shinado")
            .setContentTitle("to - do")
            .setContentText("4 todos today")
            .setStyle(inboxStyle)
            .setAutoCancel(true)
            .setDefaults(Notification.DEFAULT_ALL)
            .build();
        manager.notify(4,notification);
    }
```

图 9-16　收件箱模式 Notification

后　记

以前不明白为什么会有后记这种东西，现在估计是明白了。后记大概也都是写在校稿之后发现自己问题很多，或者又发生了一些事情让自己的看法跟刚写的时候大不一样时的有感而发吧；又或者是在全篇书面语束缚下一种解放式的心理抒发。

的确，刚开始写书时的那种妄想改变世界的兴奋感已经被交稿前几天的累感给覆盖掉了。所以当我无力地发现把前言放在最后写是一个非常大的错误时，我已经忘记了最初的感受了，想感谢的人也不觉得那么感谢(说白了就是没那么矫情了)。

不知道是哪个矫情的人说出这样矫情的话：不要忘记出发时的目的。只是现在的情境跟出发时的大不相同了，在那些梦幻泡影般的理想破裂时，我们不得不戴上无所谓的面罩来保护自己。虽然有些理想落空了，自己也没有超能力，不是 MIT 毕业的，没钱环游世界，不精通六国语言，也没有经营上市公司，但是为了表现出对现实的叛逆，我依然假装坚强地坚定自己的理想，尽管知道总有一天也会随波逐流。

前些天朋友历数了我的一堆缺点。对于这些缺点我不置可否。我想起以前打球的时候总是会碰到一些看似球技拙劣，实则得分能力很强的人对位，然后被打得体无完肤。我也只好自嘲也给自己脸上贴金一番：项羽打不过刘邦，2004 年的湖人打不赢活塞，Windows 卖得比 Mac 好。这个世界上有时候就是这样，最成功的往往不是最优雅的。但是无论世道再艰难，还是要活出雄狮般的姿态。

最后来点正能量。虽然缺点都是客观存在且明知故犯的，但是从进化角度上来看，每个人都拥有一套让自己在复杂的环境生存下去的心理机制。你之所以知错不改，那必然是因为改正这个缺点所付出的代价要比保留这个缺点而把时间花在值得你关注的地方获得的收益来得大，所以你虽然不完美，但现在的你已经是经过一系列调整之后的最好的你，而以后的你，也必然比原来的你来得完美，而你所需要做的，就是努力，努力奋斗也好，努力生活也好。

路再难，姿态要好看。